I0014681

Ethical

Hacking

The Complete Beginner's Guide to Learning Ethical Hacking

(A Comprehensive Beginner's Guide to Learn and Master Ethical Hacking)

Alice Ybarr

Published By **Zoe Lawson**

Alice Ybarr

Ethical Hacking: The Complete Beginner's Guide to Learning Ethical Hacking (A Comprehensive Beginner's Guide to Learn and Master Ethical Hacking)

ISBN 978-1-998769-60-5

Legal & Disclaimer

Table of contents

Chapter 1: Cyberlaw & ethical hacking

Before you begin to dive into the worlds of cyber-security it is essential that you understand what a Hacker is. There are many definitions of hackers. This is partly because the media has placed a spotlight on hackers and groups like Anonymous. In the 90's, hacker was a term that could be used to describe someone who was a gifted programmer capable of solving complex problems. Today, hackers are usually portrayed as criminals who steal bank account information. As you can see there is a big discrepancy. The negative term "criminal" was once a positive one.

This is why you need to break down hacker into categories and get to the heart of hacking. These are three types of hackers.

1. White Hat Hacker: They're the good guys. White-hat hackers are typically cyber-security experts that are hired to test the security of an organization and find any vulnerabilities.

This hacker will only work with the permission of the hiring party. He/she will never try to uncover confidential information.

2. Black Hat Hacker: This hacker is often called the bad guy. Another term for this type is "cracker". Black hat hackers will exploit vulnerabilities to their advantage. He is the exact opposite, or "hacker," as the media refers.

3. Gray Hat Hacker (as the name implies, this hacker works somewhere in-between the two. A gray hat hacker is similar to a white one. They will not try to illegally exploit a system vulnerability or tell others how to do it. But, they may do so without permission. Gray hat hackers don't violate the law like black hats. Instead, they use their skills to find weaknesses in systems of others and then attempt to sell the information. The legal realm is as grey as this type hacker. Gray hats claim that they do not violate the law if they are conducting security research. Companies may also hire gray hat hackers in order to test

their security. In this instance, the hacker reports all of their findings but performs questionable operations without certain individuals knowing.

This chart shows that hackers do not all behave the same. Some hacker operate in compliance with the law and ethics. Others work only for their own benefit or with malicious intent. Still others can do both. In this chapter, we will examine the meaning of ethical hackers and the rise cyberlaw. It is essential to know your future enemy, the ethics of hacking and the legal implications.

Understanding Your Enemy

Understanding how criminal hackers think is key to creating a defense. It can be dangerous to not test your defenses against all possible threats.

Over the past decade things have changed in the cyber-security sector, especially in criminal circles. For the sake of the challenge, criminal hackers would exploit vulnerabilities and hack security systems. This meant that most hackers did it just for the thrill of it. These hackers have been replaced today by financially motivated attackers, who are looking to make a lot out of illegal operations.

Apart from these profit-oriented hackers there are also politically motivated organizations that are commonly referred to hacktivists. They are often motivated by political ideas, or they use illegal and legal methods. While hackers are often operating in ethically questionable areas, it is essential to be prepared for any attack. Even if there is no financial damage done, a security breach could be devastating. Because it's hard to know their motivations, cyber-security specialists don't want to share any data.

Recognize the Risk

Hackers are often invisible to the public, making it difficult for them to be detected. Network administrators and programmers often think that security breaches like this are easy to detect. However, denial of service attacks are the most obvious. Skilled hackers know how security devices are not aware of them. It is important to understand the differences between attacks and how they work so that you can spot them immediately.

It is important to know how to prepare for an attack in order to detect the danger early. Network administrators who are familiar with hacking attempts and can spot it quickly can identify the dangers days in advance. How? They might detect a ping sweep and after a couple of days, a port scanning. These are obvious signs that someone could be preparing an attack. You can find clues in many other activities that are similar to these, which could indicate an attack days or weeks before the actual security breach is attempted.

You may argue that there are several tools that can signal such attacks to you. They can even make the decision in your place. Software can make wrong judgment calls, even though it may be more accurate than a human at making calculations. Hacking tools can also serve defense purposes. This means that both the good guys as well as the bad guys use the same tool. You need to learn how to attack and understand the mentality of attackers to be able defend.

The Ethical Hacking Process

Companies and organisations need to assess the security of their system, and what damage an attacker might cause. A penetration tester is an ethical hacker who will help them test the security of their systems. Employers can be better protected by the hacker, who simulates an attack but does not cause any damage. The attacker will use the exact same techniques and customers as well as penetration testers must understand ethical hacking. It is important to communicate all steps and activities clearly to avoid misunderstandings, and to make the whole process smoother. It doesn't matter what the purpose of the testing is. You must find common ground with your customer to make sure he or she understands exactly what you are doing. But, we must first understand the difference between vulnerability assessment and penetration test before we begin to explore the whole process.

A vulnerability assessment is designed to gather large amounts data, which will show the list of vulnerabilities and what can be done to address them. How do you get such an assessment done? An automated scanning tool can be used to scan the ports and services at any given IP address. This software can also be used to check the operating system, the patch levels of various applications, the user accounts with access to the computer, and other information. All of the results are then automatically matched against known vulnerabilities within the product's data base. While this might seem like the entire process can be automated using scanners, it is actually not. While the assessment process determines the severity and impact of vulnerabilities, it can't assess their impact. The vulnerability scan only reveals which software or portion of the operating systems is vulnerable to exploits. Therefore, penetration testing should be performed as the next step.

The penetration test will try to find vulnerabilities and get access to sensitive data. Sometimes, vulnerability scanners give false positives during invasive checks. Once the scan is confirmed, the penetration test can be used to verify the results. This process does not stop with the discovery of vulnerabilities. Once they are able to gain full access, the penetration testers can jump from one system to the next. What do we mean by "full accessibility"? The tester's goal, in essence, is to find vulnerabilities in the system to allow them to take over it. The ultimate goal of the tester is to gain the same privileges as an administrator for the environment. A penetration test will expose to both the customer and tester what a malicious hacker might do and how much damage they can cause. However, how does the penetration tester determine the impact of such an attack?

During the simulated attack the tester will be looking for the exact same treasures as an attacker. It is possible to find passwords for administrators, the CEO of the company, secret corporate documents, or files that can only be accessed by top executives. This valuable trophy shows the security flaws in the system as well as the damage it can cause to an organization. A hacker might try to explain everything, but it won't matter to people who don't know tech. To prove your case, you can reveal sensitive information to the company leaders. You are a professional. Your goal is not embarrassment to the client with the information you discover. Your role is to solve problems for others and provide value to all.

Cyberlaw

There was a time in history when the legal system and all things tech were one. The incorporation of computers into all aspects of our lives has meant that crime has found its way into cyberspace. The legal sector also had to catch up. In the past decade, these systems have become intertwined, resulting in cyberlaw. You should be careful not to let this book convince you to use your new skills for malicious intent.

Companies and organizations must worry about security and digital systems. Most governments have enacted new laws that require all organizations to respect privacy and protect information. A security breach can bring about severe financial and legal consequences. As companies seek out security advisors, cyberlaw legal assistance is also required. The legislature and the government security agencies are always updating privacy and security legislation. The cyber-security specialists and the attackers constantly fight for supremacy. This creates pressure on the legal system by keeping the security sector in a constantly changing loop.

Cyberlaw covers many aspects because this field is changing so rapidly. Cyberlaw is a complex area that involves many components. Every day, millions of people log on to their computers and phones and connect to a huge network of communication. Companies and government agencies who handle personal information of citizens face potential risks due to the internet's vastness. With that said, attackers can find any entry point they like. Cyberlaw regulates the way a company handles employee data handling, system accessibility, customer interaction, etc. However, the most important component is likely the set laws that prevent and punish unauthorized use private information. Cyber-security experts need to be familiarized with these laws as they work within their boundaries. If you work outside the law system to prevent crime, it can have serious legal consequences.

Because this book is not about cyberlaw, we won't be diving into that topic. After you are certified as an ethical hacker, it is important to familiarize yourself with the cyber security laws in your state and country. Each government has its own set regulations regarding data privacy. If you live in the US, it is important to check both the state and federal laws. Be good to others and respect the laws.

Chapter 2: Hacking Basics

Now that we have an understanding of the importance and ethics behind cyberlaw, it is time to talk about the basics you need to be able to build a strong foundation.

We will be focusing on cyber-security specialists as hackers. You will see them often in this book. Communication, however, is the key to a successful project. Once you have an understanding of the terms used in the profession, we will go into more detail about penetration testing and how to write reports.

Terminology

In this section we will discuss briefly some of the most frequently used terms in the cyber security sector. This glossary will be very useful as you work through the chapters.

1. Asset: Devices and components that are used for data processing or other information handling activities. In order to protect assets from unwanted manipulation, they must be protected from unauthorized access.

2. Exploit can be any program, bug, command, or other means that allow attackers to gain access or modify the system's data. An exploit is any activity that exploits a vulnerability in an application or operating systems. It will cause unanticipated behavior, which can be exploited by hackers to their advantage.

3. Pre-engagement. We discussed the importance of having a conversation about the process and the steps that you will follow with the client in the chapter before we did the penetration test. Pre-engagement refers back to the preparatory phase that establishes the rules of engagement.

4. Milestone: Being organized can be a key aspect of professional ethical hacker work. This is where milestones can help. They allow you to organize the process into distinct stages. Each stage is clearly labeled and assigned a starting and ending date. You can use charts and spreadsheets or other

scheduling websites to track your milestones so you are able to deliver on schedule.

5. Bot: This is a program that automates certain actions. Bots can perform repetitive tasks quicker than humans, and they can work for as long or as little as you need.

6. Brute force attacks: This attack is the easiest and most popular. It does not require the destruction of computers, as the name may suggest. A brute-force attack is actually an automated technique of combining usernames, passwords, and determining the correct combination. The entire process is done automatically, but it can take some time due to the high number of possible combinations.

7. DoS attack: This is a common attack used by beginners black hat hackers. This attack can cause a network service or server to be interrupted. It's a malicious technique to block users from accessing an online service. You might experience a DoS attack to the game servers if you're playing an online video

game. This will cause all users to be disconnected.

8. Keystroke logging: This is an easy way to capture the keystrokes of a computer and any mouse buttons. The data can be used for passwords and usernames to access sensitive information. This method is more efficient than brute force attacks in that it can quickly find log information. The keylogger is installed on the computer by sending an email with a Trojan link or via a USB flash drive.

9. Malware refers to malicious programs that attack specific software or computers. Malware may take the form Trojans or worms as well as adware, spyware and ransomware.

10. Phishing refers to a type or email fraud. In order to steal private information, the intention is to send a legitimate email.

11. Rootkit: This software is used to stealthily operate without the knowledge and consent of the user. It hides many processes running in background to give access to the computer.

It's mostly used to give someone access to their computer through a back door, while hiding it from authorized users.

12. SQL Injection: Another popular method black hat hackers use for obtaining data. This involves injecting SQL code to any software or application that is data driven. This can be an example of a statement that tells an application to send all its data directly to the attacker.

Hacking is a broad subject that will lead you to many other terms. These terms will help you get an idea of the terminology you will use.

Penetration Testing

We discussed the purpose and goals of security specialists, as well as what they are. This is the only method to find vulnerabilities in a system. You must remember, however that your motivation for the testing depends on the goals and desires of the client.

Before you can start, you must first establish the rules for engagement. It is important to establish a schedule and discuss each stage of testing. Determining the scope of the penetration test clearly will ease the process, keep your client informed, as well as protect you from any potential liability. We'll show you how this could look.

Rules of Engagement

1. Before the hacker can do anything, they must first sign a Nondisclosure Agreement. As a cyber security specialist, you want legal protection from all types of liability.

2. The next step is to determine the purpose of the penetration tests and what needs testing. Sometimes you'll need to only test one part of an organization. Also, you need to know what the expected results are.

3. Once you have determined the goals of the test and the time it will take, you need to discuss the methods you will use. The client should be informed if they don't wish to use

any of the techniques. For example, some companies or organizations don't want the hacker to attempt a denialof service attack. This must be mentioned when setting the rules. It can cause confusion, delays, or even legal problems.

4. The final rule is to agree on both the liabilities as well as the responsibilities. Ethical hackers might gain sensitive information from companies, such as bank account numbers or trade secrets. Before conducting a penetration testing, you should discuss the liabilities.

Penetration Testing in Three Steps

Once all parties have established the rules and reached an agreement, it's now time to test. There are many methods and categories of penetration testing. But for now, it is important to understand the basic steps. Understanding the steps and going through them one by one will make your job easier. Let's now see what a basic penetration tester looks like.

1. Passive Scanning is where you gather information about your target, but without actually communicating with it. Passive scanning can take place by looking at social networking sites, online databases, or even searching on Google for relevant information.

2. Active scanning: Next comes the probe of the target using specialized tools. These tools are used to map networks, sniff traffic, grab banners, and other purposes.

3. Fingerprinting: This is not physical fingerprinting. At this stage, the penetration test must identify the target's operating environment and version level, all apps and their patch level, as well any open ports and active user accounts.

4. Target Identification: After the tester has collected all information relevant to the system, the target that is most useful or the most vulnerable will be chosen.

5. Vulnerability Exploiting : This is the attack itself. Specialized attack tools can be used to

target the most vulnerable parts of the system. Some of them may succeed while others might not.

6. Privilege Elevation: This stage allows the tester to have more control over the system, as though they are the authorized users. This stage grants administrative access to the system.

7. Reporting is the final step in the penetration test. The ethical hacker should document and report all aspects. This includes exploited vulnerabilities, tools used and everything found.

There are many elements to a penetration tester, but this chapter will concentrate on the basics.

A Unethical Hacker Conducts Penetration Tests

As we've discussed before, cyber-security is a sector that requires you to understand how an unethical hacker thinks about and operates. To illustrate, we will look at a penetration testing from the perspective of a black hat and compare it to the basic ethical test.

1. Selecting a Target. The hacker may choose a target in order to get revenge, profit or simply have fun. There aren't any rules or regulations. Every action is taken for personal reasons.

2. The Intermediary. An attack is never carried out directly from the hacker's own computer. The intermediary is used to hide the identity of the attacker. The intermediary is often a victim used remotely to gain entry to the target's systems.

3. Basic Test Phases. At this stage, an unethical hacker will follow the same steps as before. Even the procedures and tools will be the exact same as for a cyber-security specialist.

4. Maintaining Access: Hackers will often set up a backdoor to gain access to the victim's systems in order to be able later. This is done

by installing rootkits and leaving behind bots in order to keep access.

How an unethical hacker uses this system is entirely dependent on their intentions. Some hackers are looking to get personal information. Others want to extort companies who have vulnerabilities in their security. Only their goal is what separates the ethical hacker from the unethical. They often use the same tools and steps to reach their goal.

Methodologies for Testing Penetration

Once you have established the rules of engagement, it is time to think about the methodology. There is no single type of test that will work for every company. Everyone has different security goals. There are many questions you need to answer before choosing the right method. There are many different types of penetration testing methodologies. Each one dictates the way a test should take place. Here is a quick explanation of some of the most widely used methods:

1. OSSTMM: This stands to "Open Source Security Testing Methodology Manual" and it's a standardized methodology for penetration test. Its purpose is to establish a common baseline for all security specialists and companies performing penetration testing. This methodology dictates what parts of a network are to be tested and how they will be performed. It also explains how to analyze the data. Because they cover almost all steps of a penetration exam, penetration tests using the OSSTMM approach are extensive but tedious. This means that such tests cannot be conducted on a daily base without the necessary human resources and budget.

2. OWASP: This stands to "Open Web Application Security Project", and it is, as the name suggests a community-driven open source methodology that is specifically designed for testing web apps. This methodology provides independent data that is not influenced or controlled by any commercial or government entity. OWASP

focuses primarily on improving security of web services and applications, while OSSTMM focuses on network security. This methodology can be used by software developers, security professionals, and businesses to increase their defenses.

3. CHECK! This penetration testing methodology was born from the need to secure government networks. Governments are responsible for sensitive and classified information. This is why it is important to have consistent testing. The CHECK approach focuses primarily upon the security of data stored on a server. The CHECK methodology uses penetration testing to determine the security of the data, and how easily it can be compromised.

4. NIST is a comprehensive penetration test methodology that, unlike OSSTMM can be applied consistently and in short periods. NIST is a four-step process that involves planning, discovery and attacking as well as reporting. The planning phase begins the test, which is

where the rules and engagement are discussed. The discovery phase can then be broken down into 2 segments. The first is the basic information gathering, network scans, service detection and service detection. While the second is all about vulnerability assessment. The third stage, the attack, is the most important step of the test. This is to attempt to compromise the targeted device. Reporting is done after both attack and discovery.

Categories of penetration tests

There are three types: White Box or Black Box, Gray Box or both, depending on what the company/organization wants to test.

1. White Box: If all relevant data regarding the target are already available, then the penetration test can be considered white box. If the test is a network one, the data can include information on all versions of the applications as well the operating system. The source code will be available to the tester in order for them to do a thorough analysis. This

type testing, which requires so much information upfront, is typically only done onsite.

2. Black Box is: You might have done it yourself. This is the opposite to a whitebox penetration test. Network testing does not require any information regarding the operating system, applications or other details. So that testing can take place, only IP addresses are available. Web application testing code is not available. Black box penetration testing usually takes place externally. Most of the information is therefore kept hidden.

3. Gray Box: This penetration category is somewhere in the middle of white and black boxes testing. While some data may be provided, some information will remain hidden from the tester. For web applications, however, the source code of the application is not available. However, information on back end servers and test accounts are often given.

Writing Reports

A report is key to any successful penetration test. Being a proficient ethical hacker means knowing how to write, format, and present a report. A report should be well-organized and clear. It should also be simple to read. You also need to think about the way it is presented. You should have red headers. Your report should be easy to read and consistent. You should follow the conventions of writing and avoid grammar errors. However, you must maintain your voice throughout the text.

These tips may seem trivial and unimportant to a hacker. However, professional reports written well will increase your credibility.

You should think about your audience

Write your report with your audience in view. There are three major audience types: the executive class and management classes, as well as the technical class. Keep in mind which portion of your report is going to be the most important for each audience. While a manager may not be interested in what exploits you used to get control of a computer system, the tech department at the company will. Let's now briefly talk about each audience to see what interests them.

1. The executive audience is made up of CEOs. They will concentrate on the executive summary as well as the remediation and findings reports. Keep in mind that executives often don't have the technical knowledge necessary to read most of your report. This means that you must write your summaries with executives in mind.

2. The management audience. They will be curious about what vulnerabilities you found and how they can be improved. They are responsible for setting the security policy in an organization or company.

3. The technical audience: Your report will be of interest to both the software developers and the manager of security. They are responsible in fixing security vulnerabilities and patching them up. So they will want to review the technical details of your report.

29

For them to be able to resolve the problem, make sure you include screenshots in your report.

Report Structure

We've discussed what makes a great report and why it is so important to make it relevant for every audience. We will now examine the structure of a penetrating test report in order to better understand what type of information it should contain.

1. Cover page: This is the first page we start with. This section should include the company logo, title and brief description of test. It is important that the cover page looks professional as it will affect how customers see you.

2. Table of contents. This section is fairly self-explanatory. A clear index should be created right after the cover. This will allow each reader to jump to the appropriate section.

3. Executive summary: This the most important section of the document and can

make or destroy the entire document. The executive summary was written for CEOs or other executive leaders. It should be concise and easy to understand for those with limited technical knowledge. It is important to first define the purpose and method of the test. Next, clearly and concisely explain your results. The summary should detail the vulnerabilities and weaknesses discovered. Write down the risks discovered after your thorough analysis. Then, discuss how to lower them using the best countermeasures.

4. The remediation plan: This section of the test report is primarily for management. However, the executive class may also be interested. Remember that both audiences might not have the same technical knowledge. What does the remediation document actually do? It should include all the recommendations that you think will improve security. For example, you might recommend the installation of a new firewall or intrusion detection system. You should clearly list and describe each item.

5. The findings summary: Also known by the vulnerability assessment summaries. The findings summary: Also known as the vulnerability assessment summaries. This section contains all your findings. This section should contain charts and other visual representations that will help the audience grasp the situation. List the vulnerabilities you found and assign severity.

6. Risk assessment: The section in which you present the risks, based upon your findings. Your description should include the possible impact on the system of each vulnerability and the frequency that it may occur.

7. Methodology: You have already discussed the different methods for penetration testing. This section will provide details about the method used. This section of the report is optional, unless the client asked you to follow a particular method. In this case, you would need to detail the steps and include a flowchart showing the process.

8. This part of your report is for technical audiences. This is where you discuss your findings in depth. You'll include details about the vulnerabilities that you found, their causes, potential risks, and your recommendations to improve data security. It is important that both the security manager as well as developers know the source of the vulnerability and the steps required to fix it.

Now that you've understood the concept of penetration testing you can move forward to learning the technical aspects of hacking.

Chapter 3: Linux Basics

Linux is the best operating system for professional ethical hackers. Why Linux and not Windows? Linux is a system that gives users complete control.

Linux is an Open Source Operating System. This means you have full access and control over its source code, which you can then modify as you like. Linux has every component under its hood, and unlike Windows, you have total control. It is crucial to hacker that you have such control. Microsoft and Apple are not able to provide this level of freedom. They conceal the inner workings, as well as prevent users from accessing many components. Also, Windows can only be controlled by Microsoft, but Linux can be managed down to the last code line.

Linux is also a good choice for penetration testing. Most security and hacking tools for Linux are built for it. It doesn't matter if your choice is Windows or Mac. Your options are very limited.

Kali Linux will be our operating system for this book. If you're not familiar with this operating system you should know that there are several distributions. Ubuntu is one of the most popular operating systems. This is because it is intended for personal use and is widely used in laptops. Kali, which is a hacking-oriented operating system, will be our choice in our instance.

This chapter can be skipped if you know how to install Linux. This section is designed for people who don't know much about the operating system. We will cover installing Kali, using it to navigate, and using the terminal as a command line.

Linux Installation

We are installing Kali to get all the hacking software we'll need later. While you don't necessarily have to choose this Linux distribution, it is recommended. To do this, first go to Kali's website at www.kali.org and download the operating systems. You need to make sure that Kali is the correct version for your system. You may need to download the 32 bit version if you have an older processor. The 64 bit version is for those with a newer processor. For more information about your computer processor, visit Control Panel / System and Security / System.

After downloading Kali do not click to install. Before we get into the details of virtualbox, and using virtual machines, Inexperienced Linux users are not allowed to modify their current operating system. Kali, Kali, and any other type of operating system, do not need to be uninstalled. To run multiple systems from the same machine, you can use a virtual computer.

This is a simple method that you can use without causing any problems. This method is extremely simple. The second reason is that you can run the operating system from a virtual computer and there's no risk of it causing any damage. Kali is not recommended for beginners. If you attempt to install Kali from a computer directly, you could make a fatal error that renders your computer unusable. You'll be frustrated and will waste your time attempting to reinstall the operating systems. Kali can be used on a virtual machine to cause damage by accident, or deliberately. It does not actually put your main system at risk. You can use all the tools that you need without being afraid of them breaking.

Let's get on with it.

The Setup

VirtualBox is needed to create a virtual machine on any operating system. This software allows you to install Linux and run it without having to remove the operating system. Visit their website at www.virtualbox.org. Download the appropriate version. Next, follow the steps to install it.

Once the virtualbox manager has been installed, you can start creating new virtual machines. Click on "new," name the virtual machine Kali, and choose Linux from its menu. After that, you will need to choose Debian 64bit from the dropdown menu, or Debian32bit if running a 32bit system. Next, click on the next button to assign RAM to your virtual machine.

Although there is no standard for RAM allocation, many people agree that less than 25% should be used. You should allocate 1GB of your virtual machine's RAM if your device has 4GB. The faster your virtual machine will run, the more memory you can give it. But, remember that your main operating software needs sufficient RAM to function properly. If multiple virtual machines are to be run, it is important to ensure that you have enough RAM in order to make everything work smoothly.

The next window will ask you to create a virtual hard disk. You can click the "create" button and choose between either a fixed-sized, or dynamically allocated drive. Dynamically allocated drives are best as they will not occupy any space that isn't needed. The virtual computer will take as much space as it needs. Hit the "next" button to choose how much space your virtual machine will need. The default setting for virtual machines is usually 8GB. However, that's rarely enough. If your hard disk is large enough, you should pick at least 25GB. Hit "create" to start the process.

The next step is to add Kali to your virtual machine. The "start" button should now appear in the virtual machine manager. The manager will then ask you to provide a disk picture of the operating systems that you have already downloaded. Navigate the image to your desktop and choose it. Kali Linux will launch on the virtual machine once you have clicked the "Start" button.

After Kali has been installed you will be given a few options for starting Kali. Choose the graphical setting. If you receive an error in this stage, It's likely that virtualization is not enabled in the BIOS settings of your computer. There will be slightly different BIOS settings depending upon your system. Hyper-V should also be taken into consideration if you use Windows systems. This is a rival virtualization. You can check your system to find a solution online. Next you will be asked for the language and keyboard layout. VirtualBox will then start to detect your network adapters. It can take some time, so be patient.

After the process is complete, you will be asked how to configure your network. First, set up the process to allow root access. In Linux, this is the administrator who has complete access to the system. Next, you need to partition the disk. A disk is a section of your hard-drive. Kali will automatically setup everything if you choose "Guided-use entire disk". Kali may warn you that Kali could delete your hard drives. Don't worry though, as this is a virtual disk. It is empty right now, so nothing will happen. The system will then ask you if all your files are on one partition, or if multiple partitions are desired. While a typical Linux installation would have several partitions to store your files, this virtual machine can hold all of them in the same location. After clicking "continue", choose "Finish Partitioning and Write Changes to the Disk" and click "continue." Now your operating system is installed.

After installation is complete you will be asked whether or not you want to use the network mirror. Because you won't be using one, click on "No". Next, you will be asked if "GRUB" is installed. Clicking on "yes", will give you the option to choose which system to boot when you start up. You have the option to choose to boot Kali, or any system you would like to install on a virtual device. You can choose to "enter device manually" and then select where GRUB should be installed. Just click "Next" to finish! Kali was successfully installed. It will now reboot and display the user login screen. You will need to log in with the root username and the password you created for this user. Kali's Linux Desktop will now greet you.

Congratulations on installing Linux successfully on a virtual machine.

Grand Tour

Now that Kali has been installed, you should be eager for the next step: to create, explore and break things. To make it easier to navigate Kali, here are some terms and concepts. We won't get into boredom because our aim is to get you off to a good start as soon possible. Let's take a look at Linux from a different perspective.

Common Terms

You need to be familiar with a few terms in order understand the fundamental concepts of Kali Linux. Here are the most crucial terms:

1. Binaries: These are executable files. These files are identical to executables for Windows users. These files are typically found in /usr/bin. They also include hacking apps such as wireless hacking and intrusion detection software.

2. Linux is sensitive to case: Linux is more case sensitive than Windows. What does that mean? Linux is a great example of this. The word "desktop", which can be used in Linux, means something different than "Desktop", "deskTop", and so forth. Any of these names could refer to files or directories in different places. Attention to the case. Searching for a file named "Test" and typing "test" will result in an error.

3. Script is a set of commands which are converted to source codes. Many tools for hacking can actually be described as scripts. They run through scripting language translators such as Ruby or Python. While we are on the subject, it is worth noting that Python is by far the most popular among all types.

4. Shell: This is a shell that allows Linux commands to be run. Bash is most commonly used shell. We will also use bash in this book whenever we need to refer to the shell.

5. Terminal: A command line interface for communicating with the operating-system and issuing commands through input instructions. The Linux terminal works in a similar way to Windows Command Prompt.

You are now familiar with the basics of Linux and can begin exploring Kali.

Terminal

The terminal is the most important thing to learn about Linux. If you are familiar with the right commands, you can do most things through this interface that is quite simple.

The terminal is a tool that allows you to run text-based commands. This environment is commonly known as a "shell". You should now see the icon to the terminal on the desktop after you have installed Kali. Double-click it in order to start it. It will open with a black screen on which you can type in your instructions.

We'll be covering some basics commands shortly, but first let's understand how Linux's file system works.

The File System

If you've done any research you may have noticed Kali, or Linux in general doesn't use the exact same file system as Windows and Mac. It doesn't have a C:drive nor D:drive. The structure is however logical and easy-to-understand.

"/". is your root file system. This is the root directory. Keep in mind, this is not the root we mentioned earlier. This is the main account which holds all the administrative power. Don't confuse the two because there is a root subdirectory which is actually the home directory. Here are some more important directories that you should know:

1. /etc is the directory that contains the configuration files. They control the start-up of programs and the system.

2. /home: This is a user's main directory.

3. /media: This area houses all media devices, including flash drives and DVDs.

4. /bin: Here are executables and application binaries for Windows users.

5. /lib - This directory contains libraries similar to dynamically connected libraries found on other operating platforms.

These are the directories that you will use frequently when working with Linux. To be able to use a command line terminal, you must know these directories.

All of these folders can be modified if you have the proper user privileges. You cannot make any changes to the main directory or some system files unless you are logged into as the root user, who has full administrative authority. You can create as many accounts as you wish, but only one account may be the master account. Start creating a user account for yourself immediately. You don't want root privileges to keep you logged in constantly. No matter if you're an experienced user or beginner, mistakes happen. Important files and directories cannot be deleted by other users.

You might think it would be a hassle to switch between a regular account or an admin account depending on what you need. Linux allows you to issue administration commands even though you're not logged into as a regular user. This is why the "sudo' command is so powerful. It stands as "superuser" and is used to indicate that you want to ask the terminal for the root password whenever you issue an "sudo", before any other commands. This is an excellent shortcut to use when working with any Linux system file system. However, you need to make sure that you follow the instructions. When you use "sudo", the system won't ask you a question about what you're doing. It will do the job.

Basic Terminal Commands

Now that we have an overview of the Linux file structure and are familiar with common terminal commands, we can begin to discuss them. This section is designed to get you started, so we will only cover the basics.

1. pwd. Since we don't use a visual interface like Windows or Mac, it is possible to not know which directory or subdirectory your are. To continue navigation or issue additional commands, you will need a way to locate yourself. To view your file system location, enter pwd. This will result in something like "/home". This will show that you are located within the home directory.

2. ls This command displays all files found within a directory. To illustrate, if you type "ls /etc", you'll get a list listing all the files found in the etc directory. If you add a "-l" after the ls instruction, you can view the file permissions. Some files might not be accessible to regular users.

3. cd. The command for "change directory" stands for the following: Type cd/etc to change to /etc. To confirm that the command worked, you can type pwd. This will tell the system where you are. To navigate to the root

directory where everything is stored, type cd, followed by two dots.

4. ./filename : This command allows you to execute a particular program. The only thing you need to do is specify the name. However, some programs cannot be accessed without administrative rights.

5. rm filename : Use this command for deleting a specific file or program. Remember that this is an irreversible action and cannot be reversed. Everything you delete using this method is permanent.

6. cat name: This command is used to preview a particular text file. Sometimes the file might be unclear. You should only use it for text files. If you try the command with an image, it will show you a lot random symbols and letter that don't have any meaning to you.

7. mkdir/rmdir: Create and then delete a directory. Be aware that to remove a directory using the "rmdir command", you will need to empty it first.

8. mv name newfilename: Indicate which file you wish to rename. Be aware that the older version will be permanently deleted.

9. cpfilename You can also use it to move the new copy. Here's an example of copying a file and setting its new location: cp myfile.jpg ../MyFiles/mynewfile.jpg.

10. man This keyword will open a manual that includes all relevant data to a given command. You can also browse it to discover other commands that correspond with the keyword you are looking for.

11. --help or the -h option or? If you need to know the purpose of an application or command line, but don't want a detailed manual, these help commands are for you. Nearly every program has a help page. You should be aware that some commands might not work every time. A double dash followed up by help (-help) is required to open a help file. Use any of the following commands.

12. grep. This command allows you to run a full search of all your files and directories. It is sort of like a mini search engine. For example, you could type grepmilk dairylist.txt into the program and it will search the "dairylist.txt" file looking for the line containing the word milk.

13. exit

These commands do not have to be memorized instantly. It is important to take time to open the terminal, and then start playing with them. Try new things and get creative. This is how you run Linux on a virtual computer. There are no known problems. You can reinstall your system without causing any damage to the main operating system.

Linux Networking

Most penetration testing will occur over a network. This is why it is so important to learn about networking. Aspiring hackers must be able understand how connections are made and how they interact with the network. This section covers basic Linux networking concepts and some useful tools that can be used to analyze and manage networks.

Active Networks - A Comprehensive Examination

An ethical hacker will often need to examine and analyze active network connections. It is much simpler than it seems. Just open the terminal. Type ifconfig. This command can be used to communicate with a network. After you enter this line, you'll receive information about the current network. This should look something like the following:

eth0Linkencap:EthernetHWaddr 00:0c:26:ba:81:0f

inet   addr:192.148.163.121 Bcast:192.148.163.255 Mask:255.255.255.0

lo Linkencap.Local Loopback

inet Addr:127.0.0.1. Mask:255.0.0.0

wlan0 link encap:EthernetHWaddr at 00:c0.ca:3f.ee:02

It may sound like gibberish, so let's see what this data means.

The first line of data is about the interface eth0. This stands for Ethernet. It is the first wired internet connection to be identified. How do we know that it is the first? Because the 0 represents 1. Linux, as in many programming languages, starts counting at 0 rather than 1. If we had a second wired connection, it would appear as eth1. The next piece of information, "EthernetHWaddr", informs us that we have the following Ethernet network type: "00.0c:26.ba.81:0f".

In the second line, information was received about the IP Address of the network. This is 192.148.163.121. Next is the broadcast address (Bcast), required to transmit information to all IP addresses. Next, we have to mention the network mask (Mask), that determines which part is connected to the local networks.

We have another network (lo) in the third line. This is a loopback address also known as localhost and allows you connect to your network.

The third line, which is also a type of connection (wlan0), is another type.

With the information you have gathered through the "ifconfig", you can set up your LAN settings and modify them as required. This data is critical to hacker skills.

Network Information is constantly changing

A hacker who is ethical will need to understand the basics of changing information in your network. Why? It's because changing your IP address can fool other networks into believing that you are authorized. If you need to denial service attack, it is possible to mask your IP. By masking your IP, your attack will appear to be coming from an unknown source. This can allow you to evade capture. Let's examine how this can be accomplished using the "ifconfig" command we already know.

Open the terminal, and type "ifconfig", followed closely by the network that you want to reassign and an IP address. Here's what the command looks like:

ifconfig eth0 192.137.182.114

That's it! If you have entered the command correctly, it will not return any error. To verify that your IP address has been modified, you can also use the "ifconfig command."

Next, we'll modify the broadcast address (network mask) and create a new one. For this, we will use the "ifconfig", command again. The procedure is similar to changing an IP adress. After entering the "ifconfig" command you will need to enter the network address, then the broadcast address and then the chosen network. Here is how the command should look inside the terminal.

ifconfig eth0 192.137.182.114 netmask 255.255.0.0 broadcast 192.137.1.255

Once you've entered the command you can see if there have been any changes. Enter "ifconfig" like before to see your broadcast address and network mask.

Last but not least, the "ifconfig" command can be used for changing your hardware address (HWaddr), also known under the MAC Address. This unique address is meant to keep hackers away and trace the origin of attacks. This address can be changed just like the IP address. Let's look at how you can change the MAC addresses. Start by typing "ifconfig", then follow that with the network name, followed by "down". This will immediately take down your network. The "ifconfig", command can be entered again. "hw" stands for hardware. "ether" stands for Ethernet. Once you have added the new information to the network, you can bring it back up the same as when you brought the network down. But this time, use the "up" command.

ifconfig eth0 down

ifconfig eth0 hw ether 00:11:22:33:44:55

ifconfig eth0 up

You should use "ifconfig", to verify that the MAC addresses were correctly updated to the new address.

The Domain Name System

DNS's main purpose is to translate domain names into IP addresses. This makes DNS an integral part of the Internet. Hackers could use this information to identify targets. With the "dig command, you can gather DNS data on your target domain. This information is critical to the preparation before an attack.

As we've discussed, the first step should be to gather information on your target. You can find out what information you can get using the dig command. You can get the IP address and the email server of your domain. Let's examine an example of such an action. Type in the terminal:

dig hacking-is-awesome.com ns

The nameserver can be identified by the "ns" at the end. This report will be generated.

;; QSTION SECTION

;hacking-is-awesome.com. IN NS

;;

hacking-is-awesome.com. 5 IN NS

hacking-is-awesome.com.   5   INNS   NS 6.wixdns.net

;; ADDITIONAL SENTION:

ns6.wixdns.net. 5 IN A 216.239.32.100

Let's look at the query and see how it is broken down. The question section contains the query type. In our example, it's the ns type, which means we are using the dig command to find out the name server of hacking-is-awesome.com. We get the answer to the question in the answer section. This is the nameserver for the domain. We also have the IP address to the DNS server in the Additional Section.

This is only one way to use the "dig" command. It can also be used for information about the email server associated to the domain. This is done using the "mx" command instead of the "ns" command. It stands to email exchange. The data you receive can then be used for attacks on email servers. Here's an example of how this command might look in the terminal.

dig hacking-is-awesome.com mx

The result is:

;; QSTION SECTION

; hacking-is-awesome.com. IN MX

;; AUTHORITY Section:

hacking-is-awesome.com.   5   IN   SOA
http://ns6.wixdns.net   support.wix.com
2019052826 10800

3200 504 700 2400

The question section serves the exact same purpose as the answer. But now, we have an author section rather than an answer section. This is where you will find the email server information.

How to Change Your DNS Server

Sometimes you might want your DNS server to be changed. Learning how to do so is a great way of learning networking skills. To accomplish this task, you'll need to leave the terminal behind and use a text editing program instead. You can use any text editing software. However, Leafpad is included in all Linux distributions.

Before closing the terminal, you will need to use it in order to open a specific file inside Leafpad. Type:

leafpad /etc/resolv.conf

You have just told Leafpad to start and then open the "resolv.conf." file. It is located in the "etc" directory. The contents of the file that you have opened should be as follows:

domain localdomain

Search your localdomain

Nameserver: 192.164.185.2

As you can see, the third line indicates that the server was set to a DNS host at 192.164.185.2. So how do we change it? Let's suppose you want Google's public DNS to be changed. This is done by changing the nameserver in the file. Type:

Nameserver 8.8.8.8

That's it! Save the file. Now Google's public DNS Server will translate domain names to IP addresses. This might take a little longer than usual, only a few milliseconds.

Chapter 4: Information Gathering

Once you have completed your training, it's now time to learn more about the first phase. Because hackers are so eager to get into hacking, they often overlook reconnaissance. Do not make the same mistake made by so many newcomers. You must do your research to ensure a successful penetration exam. Without it you might fail, stumble or even lose. This phase is not very technical and may not feel exciting at first. But stick to it.

If you think information gathering is a skill that you already have and you don't need to read through an entire chapter, then consider this real world scenario.

Assume for a moment you are an ethical hacker. You work at a cyber-security firm. You are approached by your boss about a new project. A company called him to check their security. Once the paperwork was complete, the whole process is set up. Your boss will tell you all about the company as well as any details they may have provided. His name is

"TestNet" and that's it. Hacking can be started with a word that you don't know. Where should you start? What are your plans? This can't be true. This is wrong! There should be more information about the company and employees. Wrong! This is the way your job will usually start. The first step is research.

It is not possible to know everything about a company's site, location, IP address and number of employees. Most often, you only have a name. For more information, it is best to look at publicly available data. Without having to connect directly with the target, there is much you can learn. There are two goals for this phase. Before doing anything else, you need to collect as much information about your target as possible. Once you have a clear picture of your target, you will be able to compile a list containing all possible IP addresses.

You, the ethical hacker are now the same as a hacker. Both need to gather as many relevant details as possible and analyze each target in detail before deciding on how to attack. You are however distinguished by one thing. It is important to adhere to all legal guidelines and limit your scope as a penetration tester. If you find a vulnerable system connected to your target, but not controlled by them, this is an example. You are aware that this vulnerability is critical to your task's success, but you don't have permission to use it. What are you supposed to do? It's best to ignore it. Black-hat hackers don't have to follow these rules, and they will attack in whatever way is possible. However, that doesn't sound fair. How then can you make your test more effective? In this situation, you can ignore any system that you are not authorized to use but will list the vulnerabilities and risk associated with it in your test report. On the basis of this information, you will make new decisions and perform new tests.

This introduction hopefully convinced you to not skip this section on research, information gathering. This phase is essential for future hackers. We'll continue to cover in this chapter information gathering methods as well as all the tools available to help you improve your process.

Information Gathering Techniques

To be a great information gatherer, you need to have a strategy. The reconnaissance phase is a crucial step in your hacker training. Knowing the steps will help you stand out from others and improve your learning. It is important to note that both active and passive recon strategies are common. We can group information gathering techniques into these two categories. Let's have a closer look.

What is active surveillance? Active reconnaissance refers to directly interfacing with the target in an effort to extract information. This information gathering is often used for learning about the target's open ports, operating systems, and services. Be aware that active recon techniques can reveal your presence to a target. Active methods can easily be detected by firewalls and may record your activity.

Passive reconnaissance, however, is quite the opposite. Your presence is not recorded or detected in logs because you do not interact with the target. Passive techniques rely on the Internet. Search engines and social media are some of the ways that research is done.

Let's look closer at how individual information is collected. Before you can start gathering information, you need to create a filing cabinet. It's easy to end up with hundreds upon pages of information. If your filing system is not organized, you may easily become overwhelmed by the volume. This is especially true when you're like old-school hackers and prefer to print out everything and record it on paper. No matter what method you use to organize your information so that you have access to the right information at any time. Even the simplest piece of text could end up being a valuable resource for your penetration testing.

Use a website copier

It is worth looking at the website of your target, even if only their name is known. The Internet is a place where everyone has a face. A quick search on the target's website will yield a wealth of data. Be aware that even if your visit is limited to a specific website, there might be traces of your activity. This is why it's important to limit how much time you spend there, or use a website-copying tool instead.

HTTrack, a website copier, creates a copy and makes it available offline. You will have full access, including photos and source codes, to everything on the website. This will allow you to analyze the website's contents for as many hours as you wish, and even for days. Thus, you will be able to spend very little time connected to the target web server.

HTTrack can be installed on your Kali Linux virtual machine easily. The tool is free to download. Open the terminal and type this command:

apt-get install httrack

This will start the installation of the program. After installation is complete you can launch the program by typing "httrack". Be aware that cloning websites is an intrusion. You could be responsible for the actions. You should only authorize website copiers. HTTrack will then guide you through the questions. While you don't have to answer the questions, you can always leave default responses. But you should review them so you don't get lost in the process. At the minimum, you will need to enter the project name and link to your website. Once you are done answering the questions, just type "Y" for the cloning to begin. While it shouldn't take much time, each website will have its own unique timeline. It will take more time to copy websites that have complex or large URLs. It is important to ensure you have ample space on your hard drives.

After the Cloning process is complete, the terminal will display a message that reads "Done." HTTrack is a great tool! If you've used the default settings, the website will now be located inside /root/websites/name. Next, start the browser of choice. Firefox will suffice for our example. Start Firefox. Type the location of the clone into your web address bar. From now on, you will be able follow any website link you like. Index.html is a good place to start.

Now that you have access, either by browsing the website directly online or copying it for offline usage, it is time to begin reviewing any information you find. Attention to detail is essential at this stage. This stage will allow you to find out the location of the company and contact information, such as email addresses and phone numbers. You can also determine business hours, partnership and collaboration details, employee names, social media accounts, and other pertinent information.

You should pay particular attention to "news" and "announcements" pages. Many organizations and companies love to show off their latest accomplishments. These stories often contain useful information that the company or organization may not be aware of. You can also find useful data in the job posting section. Why should you be concerned about job offers? They may be looking for new tech employees in some cases. The job description will often include information about their company's technologies. There will be information on both hardware and software. This information can allow you to hack into their systems, without being detected. But what happens if you don't see any job listings on the website. Browse the many websites and applications across the country where companies have posted job opportunities. Just enter the company name and you'll find extensive information about the people they are looking for as well as the tech they use.

Let's say, for example that you learn that the company needs Network Administrators with Cisco ASA expertise. What does this mean for you, and what do you need to do? You can draw some conclusions based on the data and also make educated guesses. It is clear from the job description that the company uses a Cisco ASA, an adaptive security appliance firewall. Based on information you've gathered from other websites, you should know how large the company is. If you have an idea of how many employees the company has, it's possible to conclude that they are either in search of a specialist or that their network admin is no longer able to set up and use such firewalls.

As you can see, a person's website can provide a wealth of information. The website should tell you at least what the company does, who they're, and what technologies they use. This information will allow you to continue passive reconnaissance, which will provide more detail. You don't have to interact with the target system, so this research method is almost without risk. It is important to start by doing a thorough online search, primarily Google.

Google Directives

Google is, undoubtedly, the best search engine when it comes to cataloging information from any part of the Internet. They are so proficient at what they do, some hackers can even complete the entire penetration testing with just Google. This section does not focus on that.

You may think that you are already familiar with Google. You've been doing this for years. Open a web browser and navigate to Google.com. Search for any item you like. This is not the case. It might be enough for the majority of internet users, but not enough for hackers. Optimizing your search will get you the best results. Google will help you gather information faster and more efficiently than any other search engine. How do you refine your Google search and optimize it? Directives are key!

Google directives can be defined keywords that enable users to search for more specific information. Google may only show you the first few results when you try to search for hacking courses from a certain university. With the right directive, you can make Google do your bidding, and extract information only from the university's website. In this case, we know keywords because we are searching for hacking courses, and we also know the URL of the university. We want results only from this website. This is possible by using the site: directive. It forces Google only to show results from the targeted site. This is what your search should look.

site:domain keywords

It is important to note that spaces should not be allowed between directive, colon, domain address. If you use our university hacking course example as a guide, you should now see links that take you directly to the university website. This directive is helpful because it removes thousands of results. You only need to review a small number of results for the target that you are interested. This allows you to focus your reconnaissance efforts without wasting time.

Intitle is another Google directive. It is easy to use as it will only return websites that contain a keyword within their title. This directive can also be modified to return websites that contain all of the keyword phrases as a title. This directive, which is self-explanatory, is called the allintitle directive. Here's a good example of where allintitle can help us with our reconnaissance.

Allintitle

Google allows you to search for directories and display a list. Many hackers begin their research with this directive. What if you want websites that include a specific keyword in their URL, instead of just the title? This is why the inurl Directive is useful. Here's how it can be used in your information gathering process.

inurl:admin

This search could result in administrative pages for your target's website. This can reveal valuable configuration information.

Google directives are fantastic at finding relevant information on targeted websites. But there are more valuable sources that you can use. The cache is one example. The Google cache can be used to search for the target's cache and further reduce your risk of being exposed as a hacker. The greatest benefit is the ability to retrieve information from deleted web pages using this directive. Google has a cache that stores all information. There are even files that have been deleted. You can even retrieve the source code that was used to build the site. Sometimes information is accidentally uploaded on the company's site and then quickly deleted. Imagine a network administrator creating an inventory of all computers names and IP addresses within a company and uploading that information to the company's internal website. They don't upload the data to the "real website". If Google's automated file deletion tools have sufficient time, the cache will store that file. An aspiring hacker should know the cache directive. Here's how to use it.

cache:testsite.com

You will see the live website if any of the links are clicked from a cached website. This could put you at risk of being detected. This directive can be used to modify your search so that you can go to a different page from the cached web site.

The last directive we will be discussing is the filetype directive. This is used in order to identify certain file extensions. This directive is used to search specific files at the target site. Let's say that we are searching for a particular file type. The command to do this would be:

filetype:doc

This allows you to search for links to any type of file. You can search PDF documents, PowerPoint presentations, and text files. Just use the appropriate keyword (the file's extension). Combining this directive with other directives makes it shine. Yes, you may combine them all, however you choose. Here's how:

site:myuniversitysite.com filetype:pdf

In our example, Google will return links to all PDF files that can be found on the "myuniversitysite.com" website.

As you can see, Google directives are a great way to reduce your research time. No longer will you need to search through thousands of results. You can scan them and find relevant data. Make sure you only use the relevant information.

Email addresses can be found

You can also use this tool to identify the email addresses of employees in an organization. However, it is impossible to manually perform this task. Kali is equipped with a simple Python script called "The Harvester". It can automate the process to catalog email addresses, subdomains, as well as server hosts. You should make sure it is up to date before you use it. An update to a search engines could have a negative affect on this tool just like any other automated program.

But what does an email address mean for information gathering? This is why it's important to imagine a scenario. Imagine an employee who has a problem. He posts about it somewhere on a social media or forum. He might even leave his email address. This email address can be found and used to hack into the company's systems. It is quite common for businesses match an employee's e-mail address to a username. Once you have an email address, you can use that information to create a few possible usernames. The usernames you have can be used to attempt to break into the company's systems. We'll be discussing this in the next section.

Now that you are aware of the importance of this step, let's start The Harvester. To open The Harvester, just type "the harvested" in the terminal. Before you do anything, open the terminal and navigate to the location where your tool is installed. If Kali is installed, most penetration tools can be found in the /usr/bin directory. You can now tell the harvester the following command to initiate:

./theharvester.py -dtestsite.com -l 10 -b google

The testsite.com website will be searched to find emails, subdomains, or hosts that are associated with it. Let's go over the command more to get a better understanding of how it works. The "./theharvester.py" part is used to call the tool. The "-d" section is necessary to indicate your target. The "-l", lower case L (not to be confused with 1), is required to limit the results we receive. In this example we tell the tool that it will only return 10 results. The "-b", or the "-s" option, is used to tell search engines which repository to use. In our case, we used Google. However you can choose LinkedIn, Bing, or other search engines. Let's examine the results we get by using The Harvester.

If your search was successful, then you will see a list containing email addresses and domains related to the target website. Although you will be able use these email addresses in a later stage of penetration testing, the new domains could prove useful for now. The entire information gathering process should be started as soon you find a domain or subdomain relevant to your target. Reconnaissance has a cyclical nature because there are always new targets you can find that will allow you to collect more information. Don't overlook new targets. They could be the key to your key. Although it will take longer, you can become an ethical hacker by learning how to properly research.

Whois

Whois allows for you to get basic information about your targeted site. You can find IP addresses, host numbers, and contact information about the domain's owner. This tool should already exist in the Linux version you are using. Now, let's open the terminal to start typing.

whois.com

Inside the terminal, you'll now see some information about your website. Now, you need to focus on the DNS servers. These DNS servers are only identified by their names. You can later use them to translate them into an IP address using the "host command", however, that will be discussed in the next section.

Instead of using Whois.net, visit www.whois.net. You can then search for the domain via it. Be aware that this information can sometimes be very limited. There is however a way around it. Pay close attention to the "whois Server" section. You can use the original search to query and obtain additional information. Here's an example to show you how much information you can find this way.

[whois.safenames.net]

Domain Name: testserver.com

[REGISTRANT]

Name of the Organisation: NewTester

Contact Name: Domain Admin

Address Line 1 Testing Boulevard 42

Address Line 2 Tester's Street Testerson

Testershire, City/Town

State/ Province:

Zip/Postcode: 123456

Country: Testland

Telephone: 1111 222 34521

Fax: 1111 222 34521

Email: domainsupport@newtester.com

[ADMIN]

Name of the organization: Testnames

Contact Name: Domain Admin

Address Line 1 PO Box 4200

Address Line 2

Testville

State/Province:

Zip/Postcode:

Country: Testland

Telephone: 1122 212 233311

Fax: 1122 212 233311

Email: hostmaster@testnames.com

[TECHNICAL]

International Testing

International Testing

Address Line 1 - PO Box 1200

Line 2 Address:

Testville - City/Town

State/Province:

Zip/Postcode: 420021

Country: Testland

Telephone: 1122 2562 273771

Fax: 1122 2562 273771

Email: tech@testnames.com

As you can see, knowing a domain name is not enough to gain much information. You also need to know how to use Whois. The address, phone number and email addresses of companies can be accessed. All this information should be kept because it will come in handy later when you are performing penetration testing.

Translating host name

As you have probably noticed, there are sometimes multiple host names that you find when collecting information. When this happens, you have the option to translate them into IP addresses using a special program. Kali already comes pre-installed with this tool. Now let's go to the terminal and type.

Host Target_Hostname

Let's suppose that, during our research, we located a DNS service with the hostname ns2.testhost.com. To translate this to an IP Address, we will need to use the following command:

host ns2.testhost.com

The end result should look something like the following:

ns2.testhost.com uses address 22.3344.555

Simple, isn't it? If you have an address and wish to find out the host name, you can use this command in reverse. Simply type:

Host IP_address

Extracting Data from DNS

DNS is a critical component of the Internet. Hackers and penetration testers can make a lot of money from it because of all the information it contains. This component handles the translation of domain names into IP address. While it may be simple to remember Google.com, computers prefer the

IP address. To keep us both happy, the DNS servers provide translation services between us and the computers.

Ethical hackers will need to pay attention to your target's DNS servers. This is because they have to know the IP addresses and domains of every machine on their network. You can find an entire treasure trove of IP addresses within organizations if you gain access to their DNS server. Keep in mind that the primary purpose of penetration testing is to obtain such IP addresses.

Another reason why DNS servers are important is that network administrators lack the experience and tend to ignore them. They follow the rule of "if something isn't broken, don't touch it." This means that many DNS servers go unpatched or updated and their configurations are rarely modified. You have many DNS servers in error that you can benefit from. But how can you get to these DNS servers?

First, you'll need an IP Address. Fortunately, we discovered DNS references in our earlier information gathering. For host names, you can use "host command" to convert them into IP addresses. Now you can begin to probe the DNS. A zone transfer is a way to extract information from the DNS. Numerous DNS servers can be used to balance the load. This is a benefit that many networks are taking advantage of. The zone transfer is a way for them to exchange information. A network administrator may not know how to properly configure a DNS server. To fix this, make a Zone Transfer and then copy the zone file. This contains all information about the servers, including their names, addresses, and functions.

The first step to performing a zone-transfer is to identify the DNS server for a particular domain. This is where your host tool from earlier can be useful. Type the terminal

host –t ns Testtransfer.com

The "host" command will perform DNS lookups of the specified target. This is the nameserver (t ns). You will now get the DNS server that looks something like this "nsttc1.test.master". Now you can type "nstc1.test.master" to perform a zone transfer.

host l testtransfer.com. nsttc1.test.master

The DNS Lookup Tool (host) attempts a zone transfer. (-l) The target domain is testtransfer.com.

You should keep in mind that the DNS configuration can make this not possible. You have other options if the zone transfer fails. The next section will cover this.

If Zone Transfer fails

As we've already said, zone transfers can prove to be very effective if the DNS server is not correctly configured or managed by network administrators who are less skilled. But many network administrators are knowledgeable and can prevent unauthorized users from performing zones transfers. There are options. There are many tools that can be used for interrogating DNS servers to extract the information you need. Fierce, for example, is one such tool. Kali comes pre-installed with this script in the /usr/bin directory. To execute the script, you just need to open a terminal and type this:

cd: /usr/bin/fierce

You can also navigate to the directory by using the "fierce", command, and refresh your memory. Let's get to the actual use of this tool. Type:

./fiercy.pl -dns testsec.com

Remember that the "./" can be used to run the program from its local directory. The command itself can be explained easily. The script will attempt the zone transfer from its target. Fierce will brute force the host names if the script fails. You can find additional targets this way that you can further question.

Social Engineering

Social engineering is an important topic in information gathering. This technique is based on taking advantage of weaknesses within human beings that are common in all organizations and companies. Social engineering works by manipulating employees until they reveal crucial information that is kept confidential.

Imagine you are doing a penetration test of your target. During your initial research phase, it is possible to find the contact information of a sales representative. It is natural to assume that someone in sales will respond to email or phone calls. You write an email pretending to interest in their products and/or services. You ask for more information. The email content is not relevant. The email itself is what you need. You can look at it and use different tools to extract data from it.

Let's move on to the social engineering example. You wrote the sales representative an email and received an automatic message stating that he was on vacation for the next week. With this information, you can contact the company to pretend to be an employee who has been away for a few days. You can also claim that you are abroad and don't have internet access for any reason. You won't be able to prove anything to the tech support of your company, so they'll likely reset your password for you. An email account that contains data about sales numbers, customers, and internal communication is now yours. Although this may sound absurd at first, professional hackers have a long history of manipulating people through email, phone calls, social media and emails. But, it takes confidence, knowledge about the company and the ability to adjust when things don't go your way.

Another great example in social engineering involves much less socializing and it is extremely common. Ever seen a flashdrive in an unusual place and wondered what is on it? Many people have. We all know what it is like to plug it in to see what it holds. This is why hackers often leave USB thumb drives at the office. It's easy to walk up, ask the receptionist some directions, then once you are distracted, "forget," the flash drive that is on the desk, or close by, it's possible to walk out. Someone will eventually pick up the flash drive and connect it with a computer. Hackers would use backdoor software to automatically launch whenever the drive is connected. These tools can run in the background, without you being aware. They allow you to connect to their systems remotely.

Never underestimate social engineering's power. People are human and make mistakes. Some people can be too trusting in their own information, believing it is not valuable. Be sure to verify that you have the authorization to obtain this information.

Data Mining

After you have exhausted all your information gathering options, it's time to set aside some time to go through all the data. After a thorough review, you should have a wealth of recorded information. This should provide enough information for you to explore the organization's structure and learn about the technologies they use.

It is important to keep track of everything you see, so you don't waste your time in the later stages. It is important to keep separate lists and files for IP addresses, host names, web addresses, and email addresses. Most of this information will contain non-IP data. This will require you to convert any non-IP data into IP addresses. Google and the host tool can be used to add more information on your IP address lists.

After you have deleted all unnecessary data and organized attackable IP addresses, take a moment to reflect on the scope of the penetration testing. Consider whether or not you are authorized to attack particular IP addresses. At this stage, it is a good idea to get in touch with the company to discuss how you can increase the scope or eliminate certain addresses from your list. Keep it simple until you have a limited list of IP addresses which you can attack.

Once you have learned the basics of reconnaissance, you should begin to practice them. It is important to practice, practice, and practice more in order to acquire knowledge. You can't really practice if you don't have the authorization to do a penetration exam. It is important to remember your ethical hacker credentials and refrain from actively gathering information. That is intrusive. It can also get you into trouble. Passive reconnaissance can be acceptable, as you don't need to connect to anyone's system. To gain information on any company, you can access the Internet freely.

Chapter 5: Ethical hacking: What are you talking about?

Hacker is often a negative term. Many stories have been reported about hacking attacks against large and small businesses. Hackers, like robbers and theft, can gain access to computer systems and view data they are not authorized to. Hackers may also hack into computers in order to steal sensitive or private information. This is done to gain their own advantage. The world has evolved from an isolated place to a global village. All continents can connect to each other and all computer networks are available to anyone from anywhere in the world. Millions of people around the world use computers, tablets, and mobile devices. These devices all stay connected to internet. This high-level connectivity makes them susceptible to hacking attacks. Millions of computer systems are affected every day by spyware, viruses, and other types of malware attacks. These hacking attacks slow down computers or cause them to crash.

You might think the original meaning of hacker was negative. In its original meaning, hacking was a positive word. Hackers are people who love to hack on computers and other electronics. They loved to learn about how computer systems work, and were eager to improve upon them. A hacker is simply a man who is interested in improving the speed and efficiency computers systems. Hackers changed the definition.

A Brief History on Ethical Hacking

There are many different perspectives on hacking. There is no standard definition of hacker. It all began at MIT in 1962, where skilled individuals worked in FORTRAN and hardcore programming. Although they were called geeks and hackers, they were in fact the most intelligent and smartest people. They were the original genius hackers, who now dominate the world of Information Technology. Hackers are always looking for knowledge. Ken Thompson was a Bell Labs employee who in 1969 invented UNIX. He changed the course of the industry. Dennis Ritchie in the 1970s gave the world C programming language which was intended to be paired-up with UNIX.

Hackers were defined as those who locked themselves in their rooms for long periods of time and programmed all day. Hacking was a thing back in those days. Hackers were highly skilled individuals who were hungry for knowledge and creation. The Federal Bureau of Investigation raided an organization called the "414s", and charged them 60 cybercrimes. Hacking was not a common practice in 1980s America. Hackers came in the form a small group who wanted to unlock computers networks and systems.

Hackers became more skilled in their fields and were able to find ways to exploit loopholes in different systems. Hackers went on a mission every time a protocol changed. Media houses began to label those who steal credit card data, crack web apps, and take money out of bank accounts as hackers.

However, reality was very different. Hackers are still often stereotyped as bad men (The History of Hacking).

Ethical Hacking 101

The word hacking is often referred to as a criminal act. However, ethical hackers identify weaknesses in computers and networks that could lead to data theft. They do security scans and secure your computer systems. Ethical hackers can also be called penetration testers. The difference between ethical hackers and other hackers is that ethical hackers get prior permission from the computer owners they hack.

A hacker testing a computer system will result in a more secure network or computer system. Hackers with ethical motivations not only spot problems but offer practical solutions. They can also assist businesses in implementing those solutions.

You are an ethical hacker. Your goal is to hack into a computer network in a way that is both secure and not destructive. You should not interfere with the operation or security of the computer systems. Ethical hacking requires that you have strong evidence of the vulnerability that has been discovered to present to the network or owner of the computer system you are testing. To show the owner the vulnerability, you could stage an attack to demonstrate it. The management of the company must be consulted to ensure the integrity data stored on their computer systems is not compromised. If you connect to the internet to share your friends' information, you should realize that this is a gateway to hackers who can gain access to your systems.

Types and types of hackers

Hackers are generally classified into the following.

White Hat Hackers

The second group is white hat hackers. These hackers discover ways that an operating systems can be exploited and how people can make a defense against possible attack. Ethical hackers monitor operating systems to ensure that security services remain up to date. They do this by continuously monitoring and actively looking for vulnerabilities and exploits. Ethical hackers are able to find new ways to tweak an electronic device in order to make it more efficient. They establish communities that enable them to share their knowledge in order to improve how others use their electronic gadgets.

Black Hat Hackers

Black hat hackers also go by the names crackers or criminal hackers. These are hackers who aim to gain malicious access on the computer systems of others for their selfish gains. They typically hack into computers to access data and modify, delete, or steal it for their personal use.

Grey Hat Hackers

Grey hat hackers possess the qualities of black hat hackers and white hat hackers. They exploit operating systems using both legal and illegal techniques. If a grey-hat hacker attempts to exploit an individual's operating system, he will inform the owner of the operating software about the intrusion and offer suggestions for improvement (Patterson n.d).

What are the risks your system faces?

A security risk is one that poses a risk to your computer's ability to function properly. An attack on your computer system that contains vital information can be a security risk. There are also non-physical causes such as virus attacks. In layman's terms, a threat is an attack that can allow you unauthorized access to a computer network.

Physical threats

Physical threats are scary events that can lead to damage or loss of computer systems. A physical threat can exist internally, like heat, humidity in areas that hold the hardware, fluctuations of power supply, and other damage to hardware. They can also exist externally, such as flooding, lightning strikes, or earthquakes. The third type of threat is human-like vandalism/theft of infrastructure, deliberate damage or accidental loss, and disruption. An organization needs to implement physical security measures to counter physical threats.

Non-Physical Dangers

Unsolicited threats, such as a non-physical one, can result in a serious incident that can damage your system data. Your business operations depend on the functionality and availability of your operating systems. Loss of sensitive information can result in your system being compromised. Hackers are able to hack into your system and continuously monitor your online activities. Non-physical threats are sometimes called logical threats. Spyware, malware Trojans, viruses, Denial Of

Service (DOS), attacks and phishing are all examples of logical dangers.

Ethical Hacking Rules

Ethical hacking has been growing in popularity as more companies invest heavily to secure their data. If you are interested in building a career within the ethical hacking industry, there are certain rules you should follow. Let's look at the rules.

Goals

The first thing an ethical hacker needs to do is think like an intrusion. He should also be aware that there are loopholes on networks and access points that can be used to commit hacking attacks. Also, you must be aware of the possible repercussions these security holes can cause and how an intruder manipulates them. The main goal of an ethical hacker, is to locate these unauthorized access point and address them head-on.

Planning

It is crucial to plan your testing phase. You must estimate the time required to conduct a hacking operation. The hacking attack will require you to have enough money. Then, you should note the number of employees that you need to do the job. Next, identify which networks you are most interested in testing. Next, define the time intervals for testing. The final step is to create a hacking program and share it afterward with all stakeholders including the company owner, IT manager, and any other persons involved.

Official Permission

You must get permission to hack into any computer network or system if you're putting together an ethical hacking plan. You must get permission in writing. This authorization should clearly state that the hacking attack will be conducted on a specific computer network. If anything goes wrong during testing, the approval should clearly state that the organization will be there to help you.

This is essential to provide protection against any criminal charges.

Stay ethical

Being ethical means you have to work professionally. Your hacking plan should be adhered to. Any deviation from this plan could be detrimental to your professional responsibilities. If you need to change from the approved plan you will need to obtain authorization from the authorities. For ethical working, you must respect confidentiality to safeguard the information you have gathered during hacking attacks. It is better to sign a Non-Disclosure Agreement to protect any information that you may find. The information includes all data on the computer systems as well as emails and passwords. Security testing results are also included. It is important to not accidentally leak sensitive information received during testing.

Keep Logs

The two main attributes that ethical hackers must possess are perseverance and thoroughness. You should never stop working on your computer. There is usually just a keyboard in your room and a darkscreen. Sometimes you may get stuck and need to work long hours to make up the lost time. To keep track of your work and know how much time is required, it's a good idea. Keep track of all the activities during hacking attacks. Next, keep track of everything you do on a daily base. To ensure safety, there should be another log. Filling in logs with date and time is the best way to keep them organized. It is essential that ethical hacking allows you to record facts, even if it doesn't work out. This is not a reason to feel ashamed. This only shows the strength of our system.

Privacy

During the hacking process, you'll be bombarded with tons of information. This information does not belong to you, therefore you must respect it. The information can include encryption keys and passwords. It should be kept private. A misuse of authority could damage your reputation, and even endanger your career in ethical hacking. The best way to deal with incoming information is to treat it as yours. Do you want to share your information?

Be Harmless

Your actions can have serious repercussions on an organization that hired you. Although you may have anticipated many of them, you may not have planned for some. You may violate the rights of others or do something that is not in your original plan. The temptation to alter the plan must be resisted, regardless of how provocative it might seem. Second, it is crucial to be familiar with your tools and how they can be used. It is very likely that your actions will result in a denial-of-service situation.

The Scientific Process

If you're using the scientific method, your work should be more widely accepted. These methods must be part and parcel of your scientific methodology.

Picking up a goal is the first step. Once you have quantified it, associate it with multiple access point or a file from an internal server. If you have two different scans of the network, it is likely that your hacking attack lacks consistency. For the recruiter to hire you for the job, you should have a convincing explanation.

Preparation of Reports

Writing reports is something you should know how to do well. Some hacking tests may be short while others can last for a longer time, especially for large organizations. They could last several weeks or even months. This type of situation requires that you prepare weekly reports, and then update the concerned parties. It is important to remember that the computer network owner or system administrator you are trying to breach into is probably nervous. If they don't hear from you, it is likely that they will panic. Report any high-profile vulnerability you find during the testing phase. You may include the following information in your report:

* Vulnerabilities like exploitation rates

* Breaches you come across

* Vulnerabilities, which can be exploited and/or untraceable

* Weaknesses that could put lives at danger

Reporting is important because it gives authorities in an organisation a way of determining the veracity of your activities (Ten Important Rules of Ethical Hackeding, 2014). This section will be extended as I complete the book.

You'll Need These Tools to Take You on Your Journey

Automated tools dominate the Internet. You can make use of them to grow your social network and answer your emails. To serve your online customers, you can also use bots. Hacking has experienced a kind of evolution. There are many automated tools you can use to perform security research in ways that were not possible before.

In the past, ethical Hacking was restricted to security professionals. Now anyone can scan a system for vulnerabilities and report them. Ethical hacking tools can be used to identify and fix potential security holes within a company and help to secure its computer systems and applications. There are many tools you can use for planning an ethical hacking testing. Here are the top most used tools, which can be both simple and powerful.

John the Ripper

John the Gipper is the most widely used password cracker available on the web. It is widely considered to be the best security tool to test the strength of the passwords that you use to encrypt the operating systems. This tool is also useful for remote auditing of an operating system. This password cracker will automatically determine what type of encryption has been used for any type of password. This tool is capable of switching its password test algorithm depending on the type and complexity of passwords. John the Ripper uses bruteforce technology for cracking passwords. It can also decipher algorithms such Kerberos AFS and Hash LM (Lan manager), DES and Blowfish (Top 15 Ethical hacking tools used in Infosec professionals, n.d.).

Nmap

Nmap (Network Mapper), a free security tool for infosec professionals, allows them to audit and manage networks from remote and local hosts. It is one among the oldest security tools and has been around since 1997. It is continually updated each year. It is recognized by security professionals as one of most efficient network mappers on market. It is well-known because of its consistent and speedy delivery of results.

Nmap helps you to audit the security and integrity of your device. It allows you to launch DNS queries against multiple domains. Nmap supports Microsoft Windows OSX, Mac OSX, Linux and other operating systems (Top 15 Ethics Hacking Tools Usen by Infosec Professionals - n.d.).

Wireshark

It's an open-source and free program that allows users the ability to analyze internet traffic. It is distinguished by its sniffing technology. Wireshark's ability to detect security threats in your network is well-known. It is well-known for its ability detect different types security problems in the network. It also helps to solve various network problems. You can activate the sniffer and intercept and then analyze the results. This makes it much more fun to identify potential threats and vulnerabilities.

It lets you save analysis from offline inspections. It is equipped with an intuitive and powerful Graphical User Interface. It lets you inspect and uncompress gzip files. It supports many devices including Bluetooth, Ethernet (Top 15 Ethical hacking tools for Infosec Professionals), FDDI/Token Ring, ATM, FDDI/Token Ring, USB, and ATM.

IronWASP

IronWASP, one of the most useful tools for ethical hackers, is available. It is easy to use, portable, open-source, multi-platform and free. This makes IronWASP ideal for security professionals wanting to audit web servers and public applications. No specialized knowledge is required by ethical hackers. It also has a Graphical User Interface. A few clicks can do a complete scan. This is the ideal tool for ethical hacking beginners (Top 15 Essential Ethical Hacking Tools Use by Infosec Professionals, n.d).

Metasploit

It is an open source project that infosec professionals can also use for their penetration testing tools to discover remote weaknesses. Metasploit Framework, written in Ruby, is the most widely used result. It allows security professionals develop and execute exploits.

Maltego

It's a powerful tool that can be used to gather intelligence or for reconnaissance. It can also help to correlate names, email IDs of names, companies, organizations, and other data. This tool can be combined with another online source such as Whois data search engines, social network, and DNS records. This allows you to examine the connections between online infrastructures, such as domain servers and web pages, netblocks or IP addresses, files, and (Top 15 Ethical hacking tools used by Infosec professionals, n.d.).

Chapter 6: Ethical Hacking: Plan, Methodology, and Process

Criminal hackers fall under the umbrella of the most strategic researchers you'll encounter in the tech space. Malicious hackers require data to increase the intensity and speed of their attacks. They wait for the perfect moment in order to acquire as much valuable information as they can in a single attack launch. They wait for the right target to get as much data possible. They would examine the daily habits and online activities of their target. They meticulously plan a strategy to attack their target using all of the skills they have.

Malicious hacking attacks can be directed against one individual or groups of people at once, but the majority of the time targets are single. He or she will never attack more than one target at a time. Hackers seek out vulnerabilities in the banking systems that they can exploit to siphon off large amounts of cash. They may have access millions of dollars and personal data to wage personal attacks. Some hackers hack into landing pages to breach the security layer of websites. Some hackers hack into accounts to stay anonymous and allow them to use other services while not paying anything.

Whatever hacker's true motivation, hackers will start an attack if he or she knows they can get through security layers. You can protect valuable information by keeping it hidden from the general public. If you have to share information because of your business model, make sure you only do so with legal users.

This chapter will take you through various aspects of security. The first thing an ethical hacker should do is to step in the shoes of a malicious hacker. Understanding the mindset of your enemy can be a crucial step. I will give you an overview of ethical hackers, such as what to expect and how to proceed. To be successful with your goals you will need the most ethical hacking strategy. I will tell you which strategy is most effective and feasible for your goals. This chapter concludes on a methodology you can use to build a powerful ethical hacking program.

Motives and motivations of a Malicious Hacker

Understanding the hacker's brain and how they work is key to gaining access to your system or network. Understanding the brain of a hacker will help you gain more power. You'll also be able to see how the system works, as well as the potential damage that he/she can inflict upon the system that you are trying protect. Many people feel chills when they hear the term "hacker". This is due to the way media has penetrated people's brains. First, you need to know that a black hat hacker is trying to access your operating system and computer network to make personal gains. Many hackers today have malicious intentions and can endanger businesses if the don't look for tangible solutions. Hackers used hacking to collect information, gain knowledge, and do challenging tasks. Hackers are usually innovative and daring. They have many vulnerabilities in mind (Beaver (2004)

Hackers see the things normal people fail to notice. They are curious about the world. They love to modify codes. Most hackers don't realize there are humans on both sides of web applications and computer networks. All hackers don't expose information or exploit it. They do not seek to expose information or exploit it. Since the advent of the internet computer hackers have been around.

The Motive

Hacking can be driven by a variety of reasons, just as with any other activity. Hackers hack to prove their skills and have a love for hacking. While this may be a malicious motive, it is not harmful to the victim. Hackers have deep, dark motives to steal information and sell it. They are obsessed with their jobs and have criminal plans. They love the thrill of the adrenaline rush when they overcome obstacles and jump into computer networks.

These people love the thrill of tackling difficult tasks at high levels (Beaver (2004)

Hackers turn to their information and feel a sense of accomplishment to make it their addiction. Some hackers want to exact revenge on their victims. They just want to make their victims miserable for being treated harshly. Hackers hack because they want revenge, snatch basic Rights, blackmail and extort Money, steal Cash, kill their curiosity, vandalism or corporate espionage.

Hackers can easily steal sensitive information from victims, including credit card numbers and bank account information. However, hackers aren't always looking for financial information. A hacker could target your trade secrets, your trade destination, and your management strategy. They may steal the data and then give it away to a corporate rival for hard currency. Hackers could hack into an ordinary computer system, compromise its information and use it as a preparation for bigger attacks (Beaver (2004)

Your corporate competitor may hire a hacker who will deface your landing surface for 24hrs. This is when your rival can gain the long-awaited market edge. This could result in you missing out on key leads. Imagine that your landing pages are compromised on Cyber Monday. Because hackers benefit from the false security perception that is so prevalent around the world, they can inflict serious damage on your operating system. While we pride ourselves on our anti-virus system, we don't realize when security is compromised and how long that lasts until we discover something unexpected. One example is that hackers can take control of your operating system, email accounts, and passwords (Beaver (2004)

Hacking is much easier today than ever. The main reason is the fact that the internet has become an indispensable part of our lives. Computer networking is also growing. Computer systems provide anonymity for hackers who are skilled in disguise. Additionally, hackers are increasingly using hacking tools all over the world. There is currently no system that can guarantee that hackers will be prosecuted, or at least investigated, if caught. Everything operates on general assumptions. Most hacking incidents go unreported. Even if they get caught, they maintain that their motives are altruistic and that the only reason they found loopholes in the system was to discourage any attempts by bad guys.

Malicious hackers, regardless of their motivations, profit off the ignorance of computer users. They take advantage of their ignorance. They attack systems that users aren't able to manage. These computer systems do not have the proper protection, security, and monitoring. These systems are easy targets for hackers who can sneak under the radars of any firewalls. They then gain access to the database by bypassing authentication systems and IDSs. The majority of network administrators can't keep up with potential vulnerabilities. Administrators become overwhelmed by security issues as information systems rapidly grow. Administrators don't have a general security plan for computer systems they manage.

Hackers are able to hack slowly, so hackers don't get caught. These hackers are hard to spot because of their slow pace. Malicious hackers launch attacks most often after business hours end. They can begin their operations at night because there is not enough surveillance.

Anonymity

Hackers love to conceal themselves behind thick screens. Hackers love to keep their voices low-key so no one can recognize them. They are adept at covering up their tracks. Administrators hate them being suspicious. They either borrow or steal a dialup from a friend or former employee of an organisation. They may also be able to buy it from a disgruntled employee fired by the company and seeking revenge.

Malicious hackers are fond of using public computers at schools, libraries, coffee shops, and other places. They may also use disposable email accounts to hide their attack. Other options include the use of open email relays, zombies and servers on a victim's network. (Beaver, 2004)

What Hackers Want

Malicious hackers search the web to locate potential targets, then select the best one. The best candidate is the one that provides descriptions of what devices they have access to. Once the hacker realizes that the person has unprotected access to an organization's security layer, he/she knows exactly what to hack.

The hacker can collect these information through online research. Hackers are able to access all public biddings and subscribers as well as SEC registrations. Hackers can obtain information about every person involved in an organization they're targeting. Hackers often search for the launching dates of websites and the webmaster providing security for the organization's security systems.

Hackers are searching for accounts and devices used by individuals or companies to make online payments or purchase items. Today's world is moving fast. Everyone uses emails, smartphones and other online payment methods, including hacking. This can make it easy to steal identities and cause irreparable harm.

Some malicious hackers may come after your social media presence. Perhaps you think that hacking your Facebook account is worth it because of the value it holds. It is actually your Facebook account that can be used as a source of information by a hacker. A hacker can access your personal data, such as your passwords. It also allows them to get your phone numbers, email addresses, and postal address. Hackers consider your email IDs to be the most valuable because it contains loads information about customers, business partners, and even family members. Email IDs are a magnet for hackers. You receive emails from banks, financial consultants firms, and many other sources. A further advantage is that it is so routinely used that we don't worry about its security.

Unlocking the Ethical Hacking Process

Ethical hacking involves following a series of phases. These phases must be followed in the proper order. The process of ethical hackers starts with reconnaissance. This is when you, an ethical hacker need to use active as well passive means for gathering information. Nmap, Maltego, and Google Dorks are some of the tools used to accomplish this task.

The nest phase is about scanning the network of all target machines you want to test. Potential vulnerabilities must be identified. Nexpose (Nessus), Nmap, and Nessus can all be useful in this process. Once you have scan the system you must enter it. You need to identify the weaknesses and then attempt to exploit the system. Metasploit, the main tool you can use, is your best friend. Next is the challenge of maintaining it once you have access. A number of backdoor channels have been installed to allow hackers to access the system whenever they want. These backdoor channels need to be set up in the computer network. Metasploit can be used to achieve this purpose.

Next is the reporting of your findings. The final step is to compile a complete report that includes all of the findings. You must mention which tools were used during the job. If you have made any illegal activity clear of your tracks, it should be mentioned in the report. Mention both the success rate and vulnerabilities of each tool during the test. The ethical hacking process is difficult to define.

How to make ethical hacking work for you

Protecting your computer system and network is possible if you know the best places for hackers to strike. Prepare before you get attacked as an ethical hacker. Before you can begin your personal ethical hacking program, you need to plan it out. It is important to have a clear plan. All information must be in a standardized format. No matter what application you are testing or whether it is a large network, you must be very clear about your objectives. These must be clearly defined and documented. You must establish the standards and be familiar with the tools needed to do the job.

First, get approval from the operator of the network or operating system that you wish to test your ethical hacking plans. This is important so your ethical hacking plans don't get canceled mid-stream. All concerned authorities should be aware of what is happening. Your approval could be revoked if you fail to get approval from the authorities. Without prior approval you may be facing legal problems. You should also update authorities on what you intend to do and possible outcomes. To protect your business in any unexpected circumstances, professional liability insurance must be purchased from an agent if the work is done by a professional. This insurance policy, also called errors & omissions insurance can be quite expensive but worth every penny.

For authorization, you can use an internal memo from your higher management if it is necessary to test security systems within your own company. If you are performing this work for a client, you will need to sign a contract. This contract must have the authorization of the customer. To avoid wasting your time or wasting effort, it is important to get written approval immediately.

Set goals

You must have written approval for your hacking tests. Then you should set clear goals for the hacking process. It is important to establish clear goals for your hacking efforts. You can further divide it into subcategories.

The first step in your ethical hacking venture must be to set your goals. Your business objectives must be aligned with them. Next, you will need to create a specific schedule with the right starting and ending dates. You need to know when the project should begin and end. It is important to have all details in writing. To put it another way, list the goals you are trying to achieve. Your ethical hacking goals must be in line with the organization's mission. Common goals in ethical hacking are to meet federal regulations or improve the company's image. Be clear on how your ethical hacking efforts can benefit the security, general operations, and Information Technology departments of the organization which has hired you. What type information is stored in the company's systems? How will the information be protected during the testing phase?

The second goal is providing financial information to customers. Your private employees may also have access to financial information. You should be aware of how much time, effort and money you're willing to spend on the ethical hacking venture.

It is a good idea to write down what kind of deliverables are possible for the company you work for. Based on your test results, you may need to create high-level reports. You will find specific information in each report that you have gleaned through the testing phase. For example, passwords, phone numbers and any other confidential data that you have.

You must document your actions to achieve your goals once you have established them. You need to work on improving the reputation of your organization if you are trying to gain a competitive advantage by keeping current customers happy and attracting new clients. Focus on the results that your ethical hacking plan can bring to the organization. The security aspect should be covered from both the technical as well as the physical perspective.

As an ethical hacker, you have the responsibility of deciding if you need help from others or whether you are able to do it yourself. Is it your intention to inform the customers of what you are doing? The customers are the people who purchase the products or services you offer. If you work at a retail outlet with more than 50 branches, you might feel the need to inform your customers about a security test. If done properly, it can enhance the brand's image among the customers. Customers may feel more protected and respected if you take them into confidence, rather than testing the systems in secret. You can inform them by email about your plans to assess the security of their systems and to let them know that there is no risk of their data being exposed. You can also publish an announcement on the website or write letters to them, to show that you are responsible and build your credibility.

It is important to include in your goals how much money you have allocated for the testing phase. You can either subscribe to online tools or purchase them. These tools are useful for testing and hacking, but you will not be able to use them if you don't have enough money. You can calculate all possible costs in advance and then give the information to the management. The latter will make sure that funds continue flowing while you test the system.

Categorize your Systems

You may want to assess your computer systems' security simultaneously or in phases depending on the severity of vulnerabilities. You can sub-divide your projects into software, operating systems, computer networks, and other areas. You can also divide them according to risk levels, such as low, medium, or high-risk.

Make sure you have a notebook handy and mark a piece of paper with "High-Risk Projects" Next, identify high-risk projects and note them in your notebook. Next, you will need to analyze which projects have medium risk. Write down any projects that pose low risk. You can also define which systems are most vulnerable or not being managed properly. This will allow you to clarify your priorities and the timeline that you must follow. Once you know what your goals are, you can create a system. This will enable you to determine your scope and calculate the resources and time required to accomplish the job.

The ethical hacker should check firewalls and routers to make sure the internet is available throughout the building. They should also check the network infrastructure. This includes email servers as well as wireless access points, printer servers, mobile devices that contain confidential information, client operating system, and client applications such as email systems.

There are many factors that influence which type of system you should test. It's easier to test all the possible aspects of a smaller network. You can test everything, from A-Z. This will allow you to test everything, from A to Z. It is possible that the company already has this information from a security review.

Timing is Important

Timing is extremely important when testing a computer network. If you test during peak hours, it will do more harm than good to the organization. It is important that you choose times when customers are less likely to interact with the network. It is best to perform a security scan or test at night or early morning hours, when customers will be less likely to visit the website. You will never know what happens during the testing phase. Unconsciously you could create a DoS-attack situation that causes customers to lose their data, which could lead to reputational damage.

Consider the customer base when creating an e-commerce site. Where are the majority of customers located? What are their business hours? There is a huge time difference between eastern and western countries.

Locate your Location

Your main goal is to hack systems from locations where malicious hackers are able to break into them. It is obvious that this is impossible to predict. It is impossible to predict exactly where they will attack. However, it is possible to make an educated guess. You can probably predict whether the hacker intends to attack the system either from the inside or outside. This is why you should cover all bases. It is possible to combine internal and external tests.

As a malicious hacker you can perform various types of attacks such as password cracking or network assessments from home.

The Tools

It is vital to use the correct tools when you are an ethical hacker. The test you are running will determine which tools you should use. It's possible to hack a test with just one tool or multiple tools. It depends on the amount of tools needed. To do comprehensive testing, you will need more tools than what you can buy online. You need the right tools to do the job. A general port scanner, such as Nmap, SuperScan or SuperScan, may be used to test passwords.

But that will not work. John the Ripper (LC4) and pwdump are better options. WebInspect, or Nikoto, is a better choice than Ethereal (Beaver 2004).

You need to know the functions of each hacking tool. You need to be able to identify the purpose of each option. It is a good idea to read through the manual before you use any tool that could cause disruption in your network. If you're unable to comprehend the functions of a tool, you can search online. For more information on a particular tool, visit Quora or other message boards like Quora.

Hacking tools may prove to be a significant threat to your network's security. That is why it is important to exercise caution when using these tools and ensure that you take all precautions necessary. You should test all options on a test server before you apply them to the target network. It's the only way to know which option will put you in a DoS scenario and cause data loss. You need to be familiar with the characteristics of hacking tools before you purchase them. Make sure you are clear about the vulnerability the tool is dealing with. Also, how to exploit them.

How to put the Plan into action

You have created a plan. Now it's time to implement it. Your first step is to prepare the groundwork for your ethical hackers attack. Ethical hacking is no longer a manual process. It has become more automated. The latest tools are available to help you perform difficult tests. Ethical Hacking can be described as beta-testing software. You need to think like a programmer. The programmer creates programs, but an ethical hacker takes them apart and interacts with their network components. As an ethical hacker you must gather pieces of information from the webspace and other sources. You then move on to understanding the system and exploring its weaknesses.

To keep track of what you have done well, and what went wrong, when you begin to work on your ethics hacking projects. You can keep track of all the tests you have run and see why they worked.

Locate Data Pockets

To determine if information about the organization is publically available, you will need to launch a reconnaissance mission. While there are a lot of information online that can be accessed by anyone, most organizations don't know about it. The systems can be viewed by anyone in the world. The process of locating information sources is known as footprinting. Let's take a look at how you can gather data.

For information about the organization, you can search the webspace using a browser. There are many resources on the internet that can provide information about the company. It could be a directory of websites or an inventory. You can conduct scans on networks and probing ports to determine what information might be vulnerable.

They may claim to have limited information about the company. However, there are many passive means by which you can obtain information through the internet. You should know what others know about and what they know about yourself. A website for an organization can contain a lot of information that might seem special to the layman. However, hackers can access a wealth of data. Websites often include information such as the names and addresses of employees, as well their phone numbers, landline numbers and email IDs. You can find important dates about different events. SEC filings can be displayed as well as press releases about the launching of new products or events. Podcasts, blogs, or presentations can provide lots of information to inform readers and generate leads. Malicious hackers could use all of this information to make your life very difficult.

Chapter 7: Physical Security and Ethical Hacking.

People tend to view hacking only as an attack on online networks. They ignore the physical effects of malicious hacking, and focus only on logical security. Physical attacks are an everyday reality. It's possible to access an organization's system by using clever techniques. Information security is dependent upon the non-technical policies adopted by an organization. An organization's physical security is the protection and management of its non-technical and technical components. It is almost always overlooked in programs. Because ethical hackers are not aware of it, it is essential for ethical hacking. No matter how much you do to secure your online spaces, if the physical space is not secured, it will be a huge challenge. The site's security is key to ethical hacking. This chapter will help you identify any weaknesses in the organization's physical security. These weaknesses will be highlighted and the best ways to fix them.

Physical Security Attacks: The Potential

Most security experts believe that security professionals are safe and secure provided they have adequately scanned networks and properly protected them. They fail to recognize that physical environment could also be a weakness. An attacker may gain access to a facility and cause a physical attack on the system. Your security systems might be compromised if hackers gain entry to a building.

Simply put, physical security requires that you secure the building from where you run your computer network. There may be laptops and hard drives in the building, as well mobile storage devices, servers, computers, or machines within an organization. Understanding the security requirements, understanding the importance and future threats to fire safety programs, and describing the various components of detection and response systems are some of the main objectives of physical protection.

Hackers Could Exploit Security Loopholes

These types of security problems may not seem to be a problem for small companies. These security concerns may be related to the size of the building, how many employees are employed, the number and location of field offices, site locations, entrance and exit points, as well the location of the server or confidential room within a physical facility.

Ethical hackers must address many security holes. The bad guys want to exploit them. First, you need to discover these weaknesses before you can develop a plan to eradicate them from your organization. Here is a list of potential weaknesses that you should keep in mind.

* If there is no receptionist in the office, that means that no one is available to guard the main door. No record will be kept of who enters or stays in the building. Which part of the identity is it? What purpose did they visit the organization?

* Publicly available computer rooms are another source of vulnerability. That is the

scenario you may have seen in movies where someone sneaks into an office building to play with the computers. He uses email to send important information and to infect systems with viruses.

* Some offices don't require visitors to get badges. They don't require that visitors sign-in at security and receive a security card.

Your staff may have a tendency to dump CDs and other portable storage devices in the trash bins. This could compromise your security.

* If there are no access control devices on the doors, this is also a serious vulnerability.

* Your office is at risk if it does not have visitor protocols. Your employees shouldn't trust anyone who visits your office to use your printer or photocopier.

* It is possible for your security to be compromised if your computer hardware or your software are not secured.

Hackers may be able to access the system by exploiting those vulnerabilities. You have to look at the system through the eyes and ears of a malicious hacker. You must be aware of any potential weaknesses in your environment. These exploits may seem unlikely, but they could open the door to hackers. Your office's layout and the design of the buildings can all be considered physical weaknesses. Physical security also involves the weather patterns of a place such as how often there are flooding, rains or hail patterns and earthquakes. It also includes crime rates in the area such as burglary or robbery.

It is vital that you consider the structure of your building when creating your ethical hacking plan. Nearly every building has doors, walls, or windows. The windows in offices are usually larger to allow more sunlight to flood the space. Malicious hackers have many options. They can exploit numerous infrastructure vulnerabilities. You need to inspect the doors of the confidential rooms for openings that could be used by an attacker to insert a device and trip the sensors.

By pushing the doors with additional power, you can verify that they can be opened. Consider the materials used in building the office doors. Steel doors are more durable than wood doors. Important files will require stronger doors. Verify if the entry points have alarm systems such as doors or windows.

How to Handle Them

By simply taking simple countermeasures, you can combat these physical security precautions. For these security measures to be countered, it is possible to consult construction professionals. For a more detailed analysis, you might hire an expert. The design should be analyzed and evaluated to determine if it will improve security.

* You have been hired by an organization to do a security screening during construction.

* You work for an organization as a security consultant and oversee the construction of a brand new office building.

Strong steel locks and doors can be recommended for installation in buildings. If the rooms will have computers, it is possible to recommend windowless walls. A security system for the office building should be recommended. It should be accessible from all entrances. It is important to ensure that proper lighting is provided at all exits and entrances. Most office buildings are equipped with razor and barbed wires.

The Layout

Pay attention to how the office layout is laid out. Hackers have the ability to exploit vulnerabilities in the design and layouts of buildings. If confidential information is on the desks of employees, you should be careful. The way your office handles the packages is also important. A malicious hacker could get a confidential package if they are not properly addressed.

Check out trash cans, dumpsters, and other containers to see if they are easily accessible. Does the management use any shredders to cut the paper before putting it in the recycle containers? An open recycling container can pose a serious problem because anyone can view its contents. A common term used to describe a plan for malicious hacking is dumpster diving. Hackers could gain access to all the information that employees of an organisation have discarded. It's not hard to see how employees dump letters and lists, and how deadly a weapon these can be for malicious hackers. You may find information about new products, information from partners or customers, as well as memos and phone lists. You might be unaware that what is most important to you may actually be of great value to hackers. A hacker can't dive into a heap of trash in front his or her entire staff.

Your copy and mail rooms should not be exposed. If hackers are able to gain access to these areas, they can steal mails and copy the company letterhead. Then, they can use it to carry out their evil plans. You should also be careful about the location of closed circuit television (CCTV), cameras. These are used to monitor the office's activities and ensure security at night.

Managers and owners often forget about access controls. You should check what type of controls your company is using. There are many choices, such as regular keys, biometrics, and card key options. These keys should be used by the person who holds them. Some companies use keypad combinations so that multiple users can access the main entrance or a particular room.

The Solutions

You must make sure you have proposed a solution to each problem. You don't have to feel satisfied that you have secured web applications, websites and computer networks. There are viable solutions for these problems. As soon as possible, you should suggest to your company's management that they employ a receptionist or security guard who can monitor the movement of people in and out the office. The main entrance to an organization is often unattended. It is important to ensure that the entrance is always open. Ask the receptionist if they can keep a log of visitor information, such as their name, address, phone numbers, and how long they spend inside. This will help deter untrusted visitors who might try to enter the organization. You can also ask your guard to assist visitors leaving the facility. It may seem like an additional expense for your budget, but it will bring benefits to your business that far exceed the negatives.

The next step is to assess your organization's culture. If they see anyone strange in the facility, you can ask them. If they observe any strange behavior, employees should inform the manager. The office can set a policy that all visitors must be directed to one room. A good idea is to place signs at sensitive locations in your office that only employees can see. A security room is a great way to curb this kind behavior. It can be set up in your office so that security guards are always on watch for visitors through CCTV cameras.

Biometrics is not a system that can be fooled. They can put a heavy squeeze on your bank account, but they are the best if your goal is to reduce security vulnerabilities in your organization.

Why is physical penetration so important?

Hackers who are determined to hack into your security will do everything they can to inflict maximum damage and penetration. Nearly all physical attacks occur during regular business hours. It is possible for attackers to mix with employees and steal their belongings. It can be difficult to attack an organization physically. It's always best to conduct such an attack during business hours. Most hackers will likely target systems during those hours. If you plan to test the systems after-hours, make sure you get permission from the management. Otherwise, you may be faced with hostile security personnel, dogs, and even police. You may convince the authorities that you're a good guy, instead of testing the security system.

Reconnaissance

Before you attempt physical penetration, you need to thoroughly research potential targets. Reconnaissance is essential because malicious hackers are likely to be thinking about it. There are a few tools you'll need to get the job done. Google Earth, Google Maps, and Google Maps are two of the most important tools. The site can also be physically inspected by making one or two visits. To be an ethical hacker, it is important to take photographs of the site. When you plan the penetration, you will have photos that you can use. So that you are able to see the structure of your building, it is essential that you get closer to the location. During the planning phase, make a list that includes the access controls and number of security cameras.

It is important to consider the number of security personnel that the company has employed to secure the main entrance. Determine if there are any opportunities to attack the organization. You should make a list with all secondary entrances. What locks are used and how many are they? The smoker's room is an unguarded space in a building where people can gain access to other rooms or cabins. You should also consider freight elevators or other service entrances during the planning phase. Test the loading docks, and make sure that the workers have access to the inside. A weakness in the office layout could be inherent. Google earth images might have clear views of loading docks, smoker's zones and other areas. Hackers can use these images to plan their attack strategy. It is therefore crucial for an ethical hacker to understand how much information google earth images can provide to a malicious hacker.

It is essential to be able to see the inside of a facility when you are actually working there.

To do this, you need to pay attention to how many digital eyes you have on you as soon as you enter the main entrance. What number of guards are there and how often are you stopped at an exit are important. There are many methods you can use in order to verify the security of the area.

To evaluate the loading docks you can use a tape measuring device, an assistant, or a clipboard. If you want to have a better understanding of the facility, carry a bag with lunch and a laptop. After lunch, you can check your emails and eat with the ground crew. The crew will see you as an employee of the facility, who is only looking for a place to eat a piece during breaks.

Improve Your Security

Surprised to find out that hacking attacks tend to be inside of jobs? The hacking is often done by disgruntled employees, suppliers, or distributors. Here are some ways to increase security in your environment and lower the likelihood of someone hacking into it.

For security purposes, it is important to have a sign-in and camera system installed. This will allow you to monitor strangers who come into your space. If you recommend a security plan for an organisation, make sure it has an external layer to prevent unwelcome visitors. Everybody should be banned from entering your backroom, regardless of whether they are a classmate, cousin, or close friend. It is forbidden to let anyone touch your main servers, or any other type of tech. Lock the server room securely with a keycard, biometric system, or lock it with keys. Only a few trusted specialists should have access to the servers. It's always better to have a keycard system that records every person who walks through the door. It will also include the date and time they entered.

Chapter 8: Don't underestimate the dark potential of social engineering

For quite some time, social engineering was misunderstood. This has led to many different opinions about social engineering and how it works. Social engineering uses the weakest link in an organization's security system to its advantage. This weakest link can also be called its own employees. This is also called people hacking. A malicious hacker uses employees' naivety to steal information and then makes it his own for personal gain.

Hackers are able to disguise themselves as somebody else. They must hide their identity in order to access information that is not available online. Hackers can use the data they take from their victims to cause chaos in the networks they are trying. They can also steal data or erase it to cause damage to the systems. Social engineering can be described as simply lying to people to extract information. A social engineer should be:

* A good actor

* A good liar

* An expert in getting stuff at no cost

Hackers typically pretend to be someone they are not in order gain information that they would otherwise not have. Hackers are able to target innocent employees within an organization in order to fulfill their objectives. They use this information to further their evil designs. While it is tempting to confuse social engineering with physical safety, in reality they are completely separate. Let's examine some examples.

Fake vendors could come to your company to provide updates to your internet and telephonic system. An ethical hacker must test whether such an attack is likely to succeed. To update the system or test it, you can ask for the security code. Most people don't believe that vendors are hackers. They simply give away passwords to the vendor and compromise the system. The vendor can have full access to your organization's system if the password you give is the administrator password. A hacker could be able to have an extended backdoor presence in the system.

Another way to attack your organization is to pretend you are a support representative. Fake support representatives from the phone company or internet may visit your business and claim they must install a new version. You could either personally visit the business or phone an employee and convince them to install the software. Once the software has been downloaded successfully, the hacker can gain full access.

Some organizations are bigger than others. Larger organizations face greater security risks as fake employees can approach security desks to request duplicate security keys. They may claim that they have lost their keys. The system will be compromised once they have access to the keys (Beaver (2004)

Social engineers can be very skilled in persuasion, storytelling, and are also highly forceful. They can tell stories and persuade people. They can also play the uninformed employee role.

What are the different types of Social Engineering Attacks

Social engineering can take many forms. It can be extremely malicious, but it also has the potential to be friendly. While it is sometimes friendly, an ethical hacker can alert the company about a potential vulnerability. Software vendors are becoming better at creating software that is difficult to crack. You can't just enter different password combinations or pin codes to get into the software. This is what led to social engineering. Social engineering softens the target, before the attacker can take it down. Social engineering comes in when hackers are unable or unwilling to perform remote hacking. Hackers are now using social engineering in order to damage their targets and weaken them, before they invade from a faraway location.

Permeability testers, who are being used to hack security systems, are also using this tactic. They are keeping up with hackers faster than ever. Social engineering is what penetration testers do for an organization. This is why organizations often hire them to test the unconventional layer of cybersecurity. The difference between a hacker and a penetration tester is that hackers don't use the information to gain their personal benefits. They use it in order to find security loopholes within an organization.

Spies are also skilled in social engineering. Spies are skilled in this type of attack. The government of a country usually employs spy agencies to gather information. Spies obtain information for the hacking group that is back in the predatory country. Spies have a reputation for tricking people into believing they are getting information. They force people to do what they want. They never stop trying until they succeed because they are trained that way. Espionage isn't limited to governments and countries. Companies can also execute it to get into the security systems in their competition. It's another way of bringing down your competition.

Social engineers are identity thieves. They may want to access information such name, address, birth date, bank account number, birthdate and other details. It is possible to impersonate another person to commit this kind of crime. A hacker can disguise himself as an employee and steal their ID card to gain access to the facility.

Social engineering is very dangerous. Most organizations don't consider it serious. They are too concerned with protecting their online systems that they often forget about it. Social engineering is an external layer of protection that is often ignored. Social engineering does not take the place of physical security. Organisations employ guards, place fences, and purchase cross shredding equipment to ensure no information leakage from discarded papers.

A person from outside the organization might use social engineering methods. It may be difficult to gather information if you are conducting the tests against someone outside your organization. It is easy to move along the process even if you are not recognized by anyone. A malicious hacker can gain access to the information no matter how strong your firewalls, access controls, or authentication devices are. These attacks are typically performed slowly by most social engineers. Every social engineer has his or her own style. He may decide to visit the facility in person or obtain the information via telephone.

Hackers who are malicious use social engineering to gain access to other systems. They want someone else to unlock the door to the company so they don't risk being arrested.

Social engineering, also known by people hacking, can be difficult because it requires a lot of information to be retrieved from employees. The problem is that the information which ethical hackers need to access is classified and employees don't want it shared with strangers. Although you might think that it would be easy to hack ethically, this is not true. Malicious hackers are skilled in this job, and they will do anything to gain access. They also have an incredible knowledge of the organization they are targeting. They are able to offer a wide range of knowledge and expertise that gives them an advantage over other staff.

Social engineering involves manipulating people. An ethical hacker must choose the weakest employees of an organization and then secure their trust. The best malicious hackers know how to read responses from employees. Here is a list of social-engineering attacks.

Phishing

Once the social engineer has decided what information he wishes to extract from the user's data, he then begins to gather as much information about his target as possible without raising alarm. If the social engineer wants to hack into an organization's security systems, he will need a list containing the names of all the employees. He will want to find the phone numbers of employees and the details of any activities carried out by the organization. You can attack the office at any time you have all the information. An employee list can be used to target the person with the most information. You can choose a communication channel that looks the least suspicious.

There are many ways to plan a phishing attack on an organization. To request the official contacts list, you can either use a fake email or phone number. You can do a scan of the target organization's social media accounts to find out who is responsible. You can easily hire someone who will scan the internet presence of the targeted organization to get more information. Once you have this information, it is possible to launch a comprehensive and powerful phishing operation.

The best way to social engineer is to reach out to your target and pretend that their account has been compromised. This will generate a sense that there is a pressing need in the company. The organization will see you as a leader and savior. Ask them for key information such the mother's maiden name, favorite pet name, childhood friend name, date of death, and the last password that the organization used. A typical target will not be arrogant and provide all necessary information without verifying it (Patterson, N.D.).

Information fishing

Social Engineers start their plans by gathering public data about the victim. Google is the best source of information. After a while, you will be able to access all the information on the company online. Online sources such as sec.com, hoovers.com allow you to view the SEC filings of these organizations. Another source of information for social engineers is dumpster diving. As we have already discussed in the chapter Physical Security dumpster diving can offer a social engineering a source of information for no cost and with little risk. People don't watch garbage cans and there is typically no security at the location. This is an effective and efficient way to obtain information.

Although it's considered difficult by some because you need to visit the site, it is still one of the most productive and effective ways to collect information. This method allows you to access confidential information. Most employees lack the necessary training to properly dispose off letters and other sources of information such as Floppy disks and CDs. For them, a piece of paper might seem like trash. But for hackers, it can be like finding gold. Sometimes, a bad guy can find the key information that he needs in order to get into the system. The following information should be found in a trash can by an ethical hacker:

* Diagrams of networks created by an organization.

* Telephone numbers of offices in the list

* Phone numbers of employees

* Organizational charts

* Lists containing passwords that have been updated (even old passwords can be used to hack your computer).

* Minutes and notes of meetings taking place within the organization

* Different reports and spreadsheets

*List of emails

* Printing emails that contain confidential information regarding the organization

While shredding is the best technique to beat dumpster diving as I mentioned, sometimes shredding does not work in all cases. It is important to be sure you are using shredders capable of cutting paper into small pieces. Never use inexpensive shredders. It is possible to make more money in this industry. You may not want a social engineering to tape pieces of paper to rejuvenate letters you don't wish him to.

Hackers can also harvest information from employees by listening in on them while they have lunch at a nearby restaurant. They can then follow them to the airports, find a nearby place to listen or engage them one-on-1. Some people use their phones to be loud and vocal. A potential leak of information could occur if an employee is talking on the other end. You can close the door on a social engineer by listening in to what the employee is saying.

There is another way to get information. Hackers can use the dial by name feature of most voice-mail systems to access the information they need. This feature is accessible by pressing 0 when calling into the main telephone number of the company or into the random phone located at an employee's workstation. Hackers can cover the number of their call to hide their identity.

To launch such an attack they could use residential phones. You can hide your numbers from the caller ID using residential phones. *67 is the code used to hide a telephone number from the caller ID. To block the phone number, all you need to do is dial this number prior to the phone number (Beaver (2004)

Social Engineering Attacks: Methodology

Building trust is key to successful social engineering attacks. Trust can be hard to gain but very easy to lose. Nature is the foundation of trust in humans. They don't trust anything unless there is a reason. They are helpful. They can be helpful. People enjoy to share their thoughts and feelings, but they may not realize the dangers. Once they gain their trust, the bad guy will reveal all the information they should have. Trust building isn't something that takes more than a day. It is important to spend time and energy on this. You must spend time with the target. Befriend him and get information only if you feel you can win his trust.

It is important to project the personality of someone you like. Be a nice person. Do not be rude to another person. Everyone loves flattery. Be kind to the other person. Make friends with him and invite him to a meal. The common interests card is the best method to build trust. You will have greater success with someone if your interests are similar to theirs. It is important to be consistent with your target's routine. One example is that he enjoys non-fiction reading and frequents book stores. You can watch him visit the bookstore and see how many times, for how long and what books. You can even pretend to buy a book while you meet him. He will be grateful for the book he has been reading. While you're checking out, pay for the book. This is how you started your friendship with him on the foundation of common interests. Then, meet him at the book store again and pretend that it is chance. Start a conversation with your target by offering him a cup coffee at a nearby cafe. Listen to him. Let him talk. Just ask him what he does and where he is located. This is all you need to know. Meet with him at the weekend. Once you have realized that your target values you highly, it is possible to get him interested in the things you desire.

Now is the time for you to make the most of your relationship. Your job as an ethical hacker is to manipulate the customer and get them to reveal more information than they need. You can also use technology on your victim to get more information. The social engineers, the evil ones who can access classified information, can use a variety of methods to get the victims' secrets. They enhance the quality and focus of the conversations they have. The victims are able to take the time to reflect back on their statements. If victims are worried or careless while performing social engineering attacks, they might want to reconsider their approach and compare it with the following tips.

* First, they must be friendly with the target. However, this should not be too

friendly. If a social engineer is excited, the target will notice that there is something wrong in their brain. This kind of situation is something you as an ethical hacker should anticipate.

* It is acceptable to tell another person that you are an employee. You can prove your claim by using a fake ID, or any other reference. Do not bragging about the authority of the organization. Your target will be aware that you lie. Everyone knows about the authority figures of an organization.

* It is not a good idea to mention names of executives, board members, or other important people in the organization. This will only make the victim suspicious and may influence his or her behavior. This will make your victim suspicious about your intentions. This will make the target feel more reserved and unlikely to share any information.

* Confidence will be the key to pulling off successful social engineering attacks. Remember that hackers are driven and willing to go to extreme lengths to accomplish their goals. Also, be prepared to do the same. Do not frown or show anxiety through your hands or feet. Be more aware of the parts of your body that are not in your face.

* Next, refuse to disclose information to the target. Ask the target questions to convince them that you are willing to share information.

* If the target shares information with you, it is not a good idea to get into too many details. While you can ask your target for details in random situations, you should not stress that. This will make them suspicious.

* An outsider should not be privy to any information you are not authorized to share. It will certainly raise alarm.

173

* Ask normal questions and avoid asking strange questions.

* It is important to keep your written communication neat and tidy if you are communicating non-verbally with the target. It should be professionally written. An spelling error can raise alarm.

Malicious hackers never lose a trick. They will first perform a favor to someone, then ask for their return. This is where the game begins. This is one the most widely used and most effective social engineering techniques. Hackers often use reverse social engineer. Hackers will always be available to help victims of a problem. They will often rise to the occasion to help, and they may even offer their assistance. Many people fall for it, I'm sure.

You can make an impression on the target by impersonating them; social engineers could wear official office uniforms, create fake ID cards or dress like employees. It doesn't matter how the target looks, they will still be considered employees of the company. They can also call it using an outside telephone number. This is one of most common methods to exploit call-center personnel and help-desk staff. This type of situation can be easy for hackers to exploit.

Tech to Deceit

Technology can make it easier. It can be more entertaining for social engineers. The request usually comes from a computer. A spoof email address, fax number and computer name can be used to launch a social-engineering attack. With the help technology, a hacker could send email to deceive and retrieve sensitive information. This dark email often contains a link which directs the victims to obtain critical information. Emails typically include a link which directs victims to a professional-looking website that can update account information such passwords, user ids, and

Social Security Numbers. This technique is often used in spam mails. Most internet users get so many spam messages every day that they forget to keep an eye on them and start to view them as normal. They have learned to ignore them and instead focus on more important messages. Spam email often appears professional and trustworthy. They trick people into sharing information that they don't want to share. This social engineering attack can also occur when hackers get access to the security systems and create pop-ups or send messages. Hackers could use these tricks via instant messaging or cell-phone text messaging (Beaver (2004)

Malicious hackers have sent malicious emails disguised as messages from Microsoft and other well known vendors to their targets, containing a patch. Internet users may think the patch is harmless and sent by some mischievous prankster. The message is actually sent by hackers who want the user to click the patch to make it install on his or her computer. The hacker could install a Trojan horses keylogger on the victim's system, or create a backdoor into the target organization's computer system. Hackers can also use victim's computers to attack other operational systems or networks. These social engineering techniques are also used to create viruses and worms.

How to Prevent Social Engineering Attacks

There are few effective defenses against social engineer attacks. An untrained user could penetrate security systems to access a network. They have many tricks in their sleeve. As an ethical hacker you can suggest many ways that an organization can protect itself from social engineering attacks.

The first thing to do is to divide the data into different sensitivity levels. Your recommendation to the organization is that they set up IDs for employees they hire. All contractors must also have their IDs. If an employee is dismissed, you can tell the company to delete their IDs from the network. A management team may also consider setting up or resetting passwords frequently. The manager can recommend that this kind of change be scheduled on a calendar. If they are too busy, they may delegate the task to someone who you trust. Each business has its own confidential information. All of these recommendations should be implemented to strengthen the security layer within your organization. The policies should be enforceable by everyone in the company. Each employee should be informed by management about the most recent changes made to the system.

A company's best defense against the threat of social engineering is to train employees to recognize these attacks and to respond to them immediately after being launched by a malicious hacker. Training employees to be aware of user behavior may start with basic training. Next, security awareness initiatives should be implemented to prevent these hacking attacks from reaching all employees.

Hire a security specialist. This is an ethical hacker who specializes on securing organizations. While it might seem costly, it's well worth the return. You can increase your defense level against social engineering by following these tips.

Owners and management of an organization should see security awareness campaigns as a business decision that will yield great returns. Training users can be done on an ongoing basis to keep security fresh in your thoughts. An awareness workshop on social engineering can help spread awareness about potential threats to your organization. It's a great idea to link awareness programs and training workshops to your security policies.

You can outsource security training to a security expert. It's not hard to understand. An outsider will help employees to take awareness and training campaigns more seriously. They are convinced that something is happening. That's why it was necessary to outsource the security training program. Do not be afraid of making an investment. It is well worth it. Before you get started with the training and awareness programs, there are a few things you need to remember. Let's take a closer look at the tips that can help to stop your social engineering attacks and even help you defeat it over time.

* Security awareness campaigns and training programs are a viable business investment. There is more to be gained than there is.

* Keep the messages within your organization as non-technical possible. This is the best prevention measure. This means that your messages should not be sent via

email, cell phones, or phone calls. As an ethical hacker, it is important to recommend that the organization's management set communication protocols. For staff who do not follow the standard protocols, the management might impose sanctions.

* You are an ethical hacker and can recommend the initiation of incentive programs for reporting and prevention of incidents in your organization. Anyone suspecting unusual activity should report it to the organization immediately. These types of reporting should be reported to a designated person in order to reduce confusion and streamline it.

* Next, it is crucial to start training programs in your workplace on an ongoing basis to keep the concepts fresh for the staff.

The following tips are possible to include in your awareness programs.

* Your staff should be trained not to disclose any type of information unless they are authorized by the person requesting it. It is easier to identify someone in person. Telephonic conversations are more difficult. Your staff can call the person's number to confirm their identity.

* It is important that your staff are trained to not open any unsolicited emails. Sometimes a link will appear on your computer screen, which opens a page that needs updating. Don't respond to it. It is an unsolicited mail.

* All guests must be properly escorted from the building by the staff.

* They should be trained to never send or receive any file from strangers. It should be your first priority to verify identity.

* The management should ensure that no stranger is allowed to connect to any of your network jacks except for a matter of seconds. Sometimes a mere few seconds

can be enough. If the stranger can be determined social engineers, he/she may install a network analyzer or spyware on your network.

* It is essential to categorize storage devices and data within an organization. You will need to categorize information assets. Your employees must be trained on how to handle each type asset in order to prevent or reverse an untoward situation.

* Every organization should have an operational computer media destruction plan. After the hard drive, pen drive, or Floppy disk is no longer needed, it must be destroyed. You must ensure that sensitive data are handled properly and that it remains where it belongs.

Cross-shredding shredders must also be used to cut paper. It is highly recommended that you outsource the task to a document-shredding business that specializes in confidential paper records disposal.

* It is important that all employees are trained on the latest online methods of fraud. Many organizations prioritize the executives. It is important to understand that the lowest-ranking employees must be considered the top priority because they are usually the easiest targets for social engineers (Beaver (2004)).